AF718569

Tales of the Tract

The beginnings of Twin Falls, Idaho and the "Magic" Valley

Books By James Varley

Steamboats, Shoshoni, Scoundrels and Such

Brigham and the Brigadier

Lola Montez; the California Adventures
of Europe's Noted Courtesan

The Legend of Joaquin Murrieta;
California's Gold Rush Bandit.

Tales of the Tract

THE BEGINNINGS OF TWIN FALLS, IDAHO AND THE "MAGIC" VALLEY

by

JAMES F. VARLEY

SECOND EDITION, REVISED AND ENLARGED

BIG LOST RIVER PRESS
TWIN FALLS, IDAHO
2008

Copyright 2004, 2008
by JAMES F. VARLEY

All rights reserved, including the rights to
translate or reproduce this work, or parts
thereof, in any form or by any media.

Library of Congress Control Number: 2008900199
ISBN 978-0-9647747-4-2

To Susan Ash and the staff of the Twin Falls Public Library,
the finest to be found anywhere in the state.

Contents

Preface

This book was written in celebration of my home town's first hundred years of life. In April of 1904, a surveyor named John Hayes, along with two helpers, came across the Snake River from Blue Lakes ranch, climbed out of the canyon, and began laying out Main Avenue and Shoshone Street on the same 45 degree bias that exists today. A few weeks later, an energetic fellow named Robert McCollum pitched a tent on the proposed town site and—equipped only with a bunch of blueprints, a water bag, and his voice and vitality—began selling town lots and water shares in the irrigation project.

From these modest beginnings, our town has grown into a beautiful and vital place to live. With a sound economic base in agriculture, and surrounded by an uncrowded countryside, it is a stable and wholesome environment in which to raise a family. All around us, recreational opportunities abound.

The stories told herein were initially published in the Twin Falls *Times-News*, intermittently, from September, 2002 through May, 2005. I first became interested in writing on the subject while indexing several early years of the *Times-News* for the Library.

The collection is certainly not intended as a comprehensive history of either Twin Falls or the Magic Valley. Rather, I hope to have given a glimpse into what life was like at the turn of the twentieth century, when the Twin Falls irrigation project was conceived and built, and the Magic Valley first became "magic." The emphasis is on the human and humorous aspect of our history.

The first edition of this book was published, as a joint venture, by the Twin Falls Public Library Foundation and Big Lost River Press, and most of the profits from its sale accrued to the Foundation. This revised and enlarged edition includes a number of stories not in the original.

Unless otherwise noted, all photographic images contained herein are provided courtesy of the Twin Falls Public Library, to the staff of which I owe a tremendous debt of gratitude—in particular Directors Arlen Call and Susan Ash, as well as Cecil and Mareda Wright. My thanks also to the Library Foundation and its executive director, Diane Van Engelen, for helping to sponsor the original project; and to Daryl Hunt for his vibrant cover design.

Twin Falls, Idaho
2008

JAMES F. VARLEY

How Shoshone Falls Got Its Name

When Twin Falls was a-building in 1905, sometimes, of a summer night, the pioneers of the tiny village could distinctly hear the booming roar of Shoshone Falls. To them the sound was awe-inspiring. The paper reported that people who didn't know its source "stood on the streets in amazement, listening to the distant thunder and speculating as to its cause."

These early settlers thought the ability to hear the sound from miles away depended upon a special condition of humidity in the atmosphere—a light rain, just before sundown, followed by a dead calm.

We don't know for sure who the first non-Indian person was to set eyes upon the falls. The Wilson Hunt expedition of fur trappers passed through the region in 1811, apparently without having seen them.

Group after group of traders, trappers, missionaries and emigrants trekked across the sagebrush plain that was destined to become the "Magic Valley" on their way to Oregon, without mention of the cataract. Even the "pathfinder," Lt. John C. Fremont, overlooked the falls while leading an expedition in September 1843.

The first written evidence of a white man seeing the wonderful waterfall came in 1847, when Augustin Blanchet, a Canadian priest traveling to Walla Walla, mentioned it in his journal. He camped for the night of August 17 at The Cedars on Snake River—the future site of Milner Dam—and on the following day, farther downstream, he saw the falls.

Blanchet gave it the name "Canadian Waterfall," because it was known almost solely to the Canadians who had been passing it since 1834 with pack trains for the fur trading

Clarence Bisbee photo of Shoshone Falls in the spring, ca. 1907.

The Bonneville Flood

Long before man came to the Snake River plain, the great cataract known as Shoshone Falls, along with the Snake's majestic canyon, was formed by a huge flood that occurred when Lake Bonneville broke out of its basin.

The Great Salt Lake is a remnant of the old lake, which surged northward into the Snake River, and down to the ocean. Caused perhaps by an earthquake, the peak flow of the flood was about 500 times that of the river ever recorded in modern times.

outposts at Forts Boise and Hall. Thus, almost certainly, one of the company's trappers was the first white man to see the great wonder.

The name bestowed by Blanchet was short-lived. On August 14, 1849, part of a mounted rifle regiment, under Col. W. W. Loring, reached Rock Creek, after a hard march along the river and across the desolate plain. The unit was going to Oregon, having just established a military post, Cantonment Loring, near Fort Hall. After making camp, some of the troopers went fishing and soon returned laden with a fine mess of trout, which were served for supper.

The next day—a dark and smoky one due to summer fires—the soldiers traveled along Rock Creek and made camp on the south wall of Snake River canyon. From there they could hear the sound of a waterfall which, at first, they thought might be either the Little Salmon Falls, or the "subterranean river" on the north bank, that had been mentioned by Fremont.

But Loring learned the next morning from his Canadian guide, Oliver Brisbors, that nearby there was a grand waterfall, the guide had once been shown by a native.

Loring allowed Brisbors to take three men to see the

Bisbee photograph of Eagle Rock atop Shoshone Falls

falls — Lieut. A. J. Lindsay of the rifles, E. D. Denny of the navy, and George Gibbs, a writer. They rode mules eastward along the canyon's edge and, after about eight miles, the splendid view of the falls broke suddenly upon them, drawing an involuntary shout from all four men.

The mules were led down a steep trail, and picketed on a patch of grass near the top of the falls. While eating their lunch of hardtack and bacon, the men decided that the name Canadian Falls, bestowed by Blanchet, was inappropriate for such a grand American wonder. And so, "representing the sovereignty of Uncle Sam," as Gibbs reported, they "baptized the cataract by the name of the Shoshonee Falls."

Then, to commemorate the occasion, Lindsay and Gibbs climbed down the sheer lava wall to the foot of the falls, where they set fire to a pile of driftwood. Through a brilliant rainbow which spanned the base of the falls, the two men could see an eagle, which had its nest on a column of rock, wheeling through the rising spray.

After the report of Loring's expedition circulated, many emigrants traveling the Oregon Trail took the time to see the natural wonder. Usually, they camped at the crossing of Rock Creek, just south of Twin Falls, and went out from there, in the evening, to take in its everlasting beauty.

The Devil's Scuttle Hole

On October 28, 1811, while canoeing down the Snake, Wilson Hunt's men reached a place where the river compresses into a narrow channel and becomes wild.

When the second of Hunt's fourteen boats attempted to pass through this chute it was destroyed, and a *voyageur,* Antoine Clappine, was drowned.

Hunt named the place Caldron Linn, but some of his men called it the Devil's Scuttle Hole. After sending exploring parties downstream for some distance, Hunt decided the river was too risky, and he walked his men to Oregon.

"Niagara Of The West" — A Gift of the Great Spirit

Albert D. Richardson, of the *New York Tribune*, was likely the first to make the comparison between Shoshone Falls and those at Niagara in his home state. In the summer of 1865, while traveling from Salt Lake City to Boise, the journalist's stagecoach stopped overnight at Rock Creek Station. Wanting to see the great falls, he arranged to have the stage driver make a detour on the following day for that purpose.

A start was made before daylight the next morning, after the passengers had spent a restless night sleeping on a bed of hay in their coach. Much later, the journalist would wax poetic, recalling how the tall sagebrush had loaded the air with a heavy perfume; how the cloud of spray from the waterfall could be seen at a distance of six miles, and how its "eternal roar"could be heard clearly at three miles.

In the romantic fashion of Victorian-era writers, he made the claim that the Indians called the falls *Pah-chu-lak-a*—a gift of the Great Spirit.

The stage line had been established just the previous year, 1864, after Ben Holladay—who operated

View from above Shoshone Falls. From Tim O'Sullivan photo in *Harpers Magazine*, September, 1869. The article accompanying this engraving was the first to use the term "Niagara of the West."

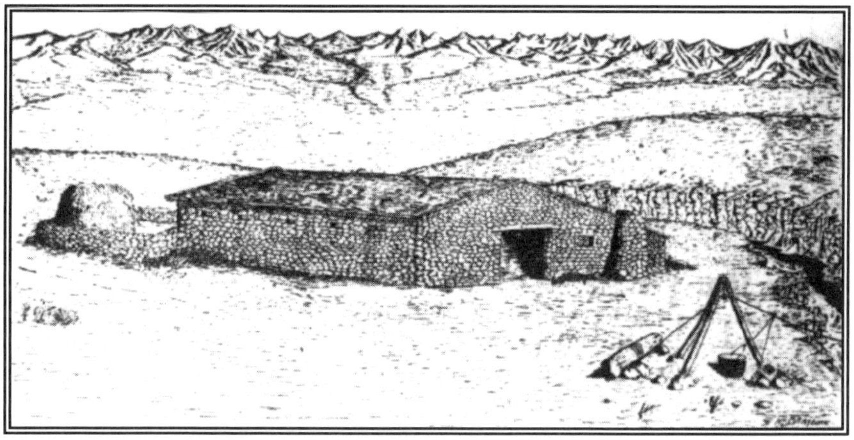

Artist H. S. Rayburn's conception of old Desert Station.

the overland line to California—was awarded a contract to carry tri-weekly mails between Salt Lake City and The Dalles, Oregon. The line was run through Boise to service the booming mining camps in Boise Basin and the Owyhee mountains.

In March, Holladay's men began establishing stations, built of lava rock, two of which were sited near what is now Twin Falls. The first—a "home station," capable of handling up to forty horses, and of providing passenger meals—was where the Oregon Trail first intersected Rock Creek, southeast of our city. The second outpost, known at first as Canyon Station, and later as Desert Station, was built atop Rock Creek canyon, just west of town.

Shortly after Holladay's first four-horse coach reached Boise on August 11, Lt. Charles Hobart was sent eastward from Fort Boise, with fifty Oregon and Washington Volunteer soldiers, to protect the stage route and the emigration from Indian attack.

A Snake River ferry had been put in that summer by Samuel F. Briggs and Samuel Clark, about twelve miles below the mouth of Rock Creek. Upon reaching this crossing, Hobart sent fifteen men, under Lt. Charles F. West, up through the rough canyon of the Snake to investigate some recent thefts of Holladay's livestock.

West and his men searched the country around Canyon and Rock Creek stations for the marauders and missing livestock, without result. They did, however, get a look at the great falls, before rejoining Hobart and returning to Boise in October.

In 1865, the year of Richardson's visit, James Bascom and John O. Corder built a general store near Rock Creek Station, the first such place between Boise and Fort Hall. That spring, soldiers were again sent out from Boise, after Indians made horse-stealing raids on several stations.

A detachment under Lt. James Currey, established a makeshift "Camp Reed" across the creek to the south of Rock Creek station, on July 18. About a week later, Indians stole four of Currey's beeves, and four horses from the station keeper, Mr. Archer. Ultimately, a group of soldiers and civilians from the Salmon Falls Ferry engaged the raiders and killed three of them.

Late in that summer, Camp Reed was moved to the Shoshone Basin, about 18 miles southwest of Rock Creek Station. There, the soldiers had some success in stopping the depredations, engaging a raiding party and recovering some stolen stock. But the next few months at Camp Reed were uneventful, and the place was vacated in 1866.

A party of U. S. Geological Survey scientists, led by Clarence King, visited Shoshone Falls in 1868, as part of a project to map an

Remnants of Camp Reed

At the time of Twin Falls' founding, the earthworks thrown up by soldiers at their camp on Shoshone Creek could still be seen, a short distance above a Shoesole Ranch corral at the timberline.

Indian bodies were also frequently found in that area—some seated upright in graves, and surrounded by personal effects, and others in the crevices of rocks, covered with timber and stones.

Charles Savage photo taken below the falls.

area extending 100 miles each side of the line along which the transcontinental railroad was then being built. King's party camped near Rock Creek the evening before reaching the falls. A fierce October wind blew all night long out of the northwest, clearing away the haze from numerous wildfires.

The dawn broke, cold, clear and calm, and, after breakfast, the explorers rode toward the canyon. As they approached it, wrote King, "a dull throbbing sound with deep pulsations" came from the ground beneath their feet.

Leaving his small cavalry escort to bring up the wagon, King and others galloped to the edge of the canyon. The party descended to the level of the river above the falls by a narrow, winding path, with photographer Timothy O'Sullivan's equipment carried upon mules.

Tents were pitched on the edge of the cliff and the expedition spent ten days camped there, during which O'Sullivan took a number of wonderful pictures on glass, whose publication gave national prominence to the falls.

But the arrival of visitors in any numbers would have to wait for the coming of the railroad. Meanwhile, it would be gold, and not the wild beauty of the view, that would attract the next wave of visitors to what was now being called the "Niagara of the West."

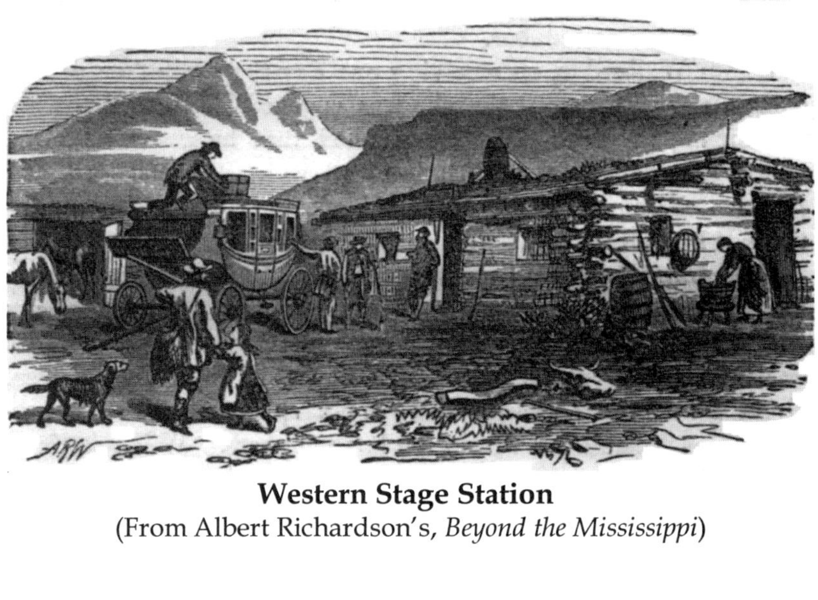

Western Stage Station
(From Albert Richardson's, *Beyond the Mississippi*)

Flour Gold

—Placer Mining in Magic Valley Didn't Pan Out—

It was a disillusioned miner who, in 1870, said, "The view of Shoshone Falls was far more rewarding than the gold prospects."

A certain Col. Kirkpatrick, and some of his friends from the Owyhee mining district, had been lured to the falls after learning that a fellow named Jamison had discovered a promising eddy the previous October. Jamison's report was believable, since he had formerly been associated with Elias Pierce, the first discoverer of gold in northern Idaho.

For a time, before leaving that autumn, Kirkpatrick's group had run five gold-separating rockers, above and below the falls, barely making a living wage. Doubtless, they had soon discovered that Snake River gold is very fine and difficult to save by the usual placer mining methods of panning or sluicing. It took 500 or more very small particles, or "colors," of this "flour gold" to equal the value of only one cent.

Placer mining near Starrh's Ferry on Snake River, ca. 1885.

Nonetheless, in that year of 1870, a modest "gold rush" began, the promise of the metal luring the first significant number of Americans to the area known today as the Magic Valley. For a few seasons, the washings were pretty good and a number of small communities sprang up along the river. Some, such as "Walterberg" and "Mudbarville," died aborning, but others fared better.

"Drytown," situated at the mouth of Dry Creek, near today's Murtaugh, grew from a single trader's tent, to include a blacksmith shop, restaurants, several saloons, and two general stores. For a time, Shoshone City, at the "Little Shoshone" (Twin) Falls, was the largest of the mining hamlets, comprising, at first, four canvas shanties and a tent—all trading posts. It later had a few saloons, a dance hall and two general stores, one on top of the canyon and one at the bottom.

But Springtown, in the canyon on the south bank, just below where the Hansen Bridge now is, eventually became the main center of commerce, perhaps because there was a flat-bottomed row boat there to ferry men and goods across the river. By the time of the gold excitement, the stage from Utah

Placer mining water wheels near Starrh's Ferry, c. 1885.
(Courtesy Shoshone Public Library)

Even Ira Perrine Was Bitten By Gold Bug

In the summer of 1884, just after arriving in Idaho, Ira Perrine and a partner ran a placer operation in Hagerman Valley, using water from Billingsley Creek to work their claim.

In 1893, he became involved with Col. Stanley B. Milner, of Salt Lake City, and others of the Milner family, in the Idaho Gold Mining Company. The company owned claims on the south side of the Snake at Jennings Bar, near Auger Falls.

Still later, in 1896 and 97, Perrine supervised the company's operations at a placer bar on the south side of the river, opposite the Blue Lakes ranch. Another man ran a similar operation for the company near Mountain Home. Each of them used a gold-saving machine called the "Gaylord Bell Mechanical Gold Pan," and heartily endorsed it's efficiency in saving the flour gold.

Nonetheless, after spending $3,000 on these gizmos, they proved a failure.

to Boise was originating at Kelton, on the line of the new transcontinental railroad, and was being operated by the North Western Stage Company, an Eastern outfit. The Rock Creek stage station—Messrs. Keene and Hyde, proprietors—became a central stopping place for the miners.

In 1870, one man warned, in a letter to the Elko *Independent*, of this "den" which was being run by "desperate characters," who, if they couldn't induce a miner to "get drunk on their tarantula juice and lose his dust at 'pick-up' cards," would "go through him by main strength and force of numbers." There was some talk amongst the miners of "getting up a hanging excursion for the benefit of the country." Later, another correspondent defended the integrity of the stage station's proprietors.

Within a few years, "gold-saving" devices came into use. These contraptions were a variation on the process used in quartz mills to attract gold particles. Copper plates were installed in rockers, sluice boxes, and other conventional placer mining devices, and these plates were coated with mercury to attract the flour gold.

When the lowest and richest of the river bars played out, hydraulic mining was used to work the high bars and terraces. It was reported that up to 1,000 tons of gravel could be processed through one device, using this method, in a 24-hour period. Water was obtained from creeks, or through ditches from the river upstream of claims. Even water from the Thousand Springs was diverted for mining.

Eventually, many of the Americans, without enough capital to start larger operations, sold or abandoned their claims to Chinese miners, who patiently gleaned from them the elusive grains of dust.

Shallow-draft suction or bucket-type dredges, with gold-saving devices, appeared on the river beginning in the 1880s and 1890s. Usually they were self-contained, and capable of doing the amalgamation, retorting and pouring into bars right on board. The Sweetser-Burroughs Mining Company built two such coal-fired, steam propelled dredges and operated them for years in the river southwest of Minidoka. In 1882, there was even such a boat dredging in the river just below Shoshone Falls.

Mining continued along the river well beyond the time Twin Falls came into existence.

Washing gold with a rocker.
(From Harper's Monthly Magazine)

"Old Dowdle Bill" – Early Days at Rock Creek Station

Charles Walgamott
(Courtesy Idaho State Historical Society)

Charles S. Walgamott, a Magic Valley pioneer, wrote several colorful histories about his experiences in the region. In one of them, he even took credit for first telling Ira Perrine about the Blue Lakes.

But there was one incident – the demise of Bill Dowdle – about which Walgamott didn't tell the whole story.

Walgamott came from Iowa to Rock Creek in August, 1875, to join his sister, Irene, and her husband, Charles Trotter, who had charge of the stage station there. To get to Rock Creek, the 17-year old had taken a train to Kelton, Utah, where he caught the North Western Stage Company's daily coach to Idaho.

The lava-rock station house, and Bascom and Corder's general store formed an oasis on a sagebrush desert, making Rock Creek a favorite stopping place for a rough clientele – freighters hauling goods from Kelton, miners working the placers on the Snake River, and a few cattlemen who were beginning to come into the area.

One morning not long after Walgamott's arrival, a certain William Dowdle—hungry and moneyless—came to the station and asked for breakfast. Charlie Trotter fed him and Dowdle left; but his demeanor had aroused the station keeper's suspicions, so he followed Dowdle and saw him mounting a horse which Trotter recognized as belonging to a friend in Boise.

Trotter arrested Dowdle, and the horse thief was subsequently convicted and imprisoned for two years. It was learned that Dowdle had previously spent a year in prison for the same offense. The prisoner vowed to kill Trotter when he was once again free.

Upon his release, in August 1877, Dowdle hired on as cook for a freighter named Johnson, and set out for Utah. Johnson and his train arrived at Rock Creek on the evening of September 14, and

Stricker Store in 1885.
(Courtesy Twin Falls Historical Society Museum)

made camp, intending to lay over for a day.

The next morning Dowdle set out to execute his threat against Trotter, but learned that the station keeper was abed, sick with typhoid fever. Frustrated, Dowdle got very drunk and then sat himself down beside the road and began firing his revolver at passers-by, slightly wounding one man.

Upon hearing shots, the young man tending the general store — which by this time was owned by Herman Stricker and John Botzet — hurried to the doorway just in time to have a bullet graze his head. He immediately grabbed his revolver from a nearby shelf and shot Dowdle through the heart. The villain fell in the road, supposedly exclaiming before he died: "Such is life, boys, in the days of forty-nine."

In his later writings, Walgamott failed to mention that he, himself, was that "young man" who gunned Dowdle down.

The inhabitants of Rock Creek unanimously approved of Walgamott's timely action, and a coroner's jury was promptly convened which exonerated him from blame. Then, the men at the station decided to bury Dowdle "with military honors," as the Boise *Statesman* reported.

A great deal of drinking was done while a rude pine coffin was built. After laying Dowdle's body in it — sheathed in freighter Johnson's overcoat — four men lifted the coffin by its rope handles and a parade of sorts began, with a crowd marching ahead of and behind the men carrying the deceased.

Since no clergyman was present to conduct services, and no

Old Rock Creek

The original settlement called Rock Creek is located five miles south and two miles east of Kimberly, and lies between the creek and the 3200 North road.

At one time, the little community consisted of the stage station, a general store, storage cellars built of lava rock, a blacksmith shop and a school.

layman was sober enough to offer a prayer or read from the Bible, the revelers decided to send Dowdle to his Maker to the music of a California Gold Rush tune entitled "The Days of Forty-Nine."

The original song had portrayed "Old Tom Moore" reminiscing about former mining comrades. One stanza commemorated "Lame Jesse," a kindred spirit to Bill Dowdle. So, the "mourners" substituted "Dowdle Bill" for the name of "Lame Jesse," and sang —

Old Dowdle Bill was a hard old case;
Who never would repent.
He never was known to miss a meal, —
Or ever pay a cent.

But old Dowdle Bill, like all the rest,
He did to Death resign;
And in his bloom went up the flume,
In the days of Forty-Nine.

The drunken revelers carried the coffin round and about Rock Creek for hours, each man simultaneously singing a different verse of the song, producing a cacophony of sound. Impetus for the parade was maintained by frequent pulls on whiskey bottles.

When a stagecoach arrived that evening from Utah, its passengers alighted just in time to see the conclusion of the bizarre ceremony. As the procession halted near the coach, the freighter, Johnson, finally realized that it was his overcoat covering Dowdle's body. Not much bothered, he lifted the corpse, removed the coat, and put it on. The march was then resumed and, with more singing and laughter, the body was borne to a low ridge and buried.

One of the travelers, an Englishman, was appalled by the wild funeral and expressed fear for his own life. He skipped the fine dinner prepared by Mrs. Trotter and — so it was said — his appetite didn't return until the next home station was reached.

The Hansens Of Rock Creek

— An Early Marriage in the "Land of Eternal Sunshine" —

John F. Hansen, 1872
(Courtesy Robert Maddock, Jr.)

We have James Iverson of Rock Creek to thank for bringing the Hansen clan to Idaho."This is the land of eternal sunshine," he wrote to a Danish newspaper in 1876, saying that it offered "opportunity for all in a health-giving climate, unequaled anywhere."

John and Lawrence Hansen — half-brothers — had emigrated from Denmark to Indianapolis, Indiana a few years earlier. Lawrence was nine years older than John and a cabinet maker by trade. He was also a widower with two young girls to raise, so he lost little time in finding them a mother. At a meeting of a Danish singing society he and John met two sisters, Anna and Marie Peterson. Lawrence married Marie in 1874, and John became engaged to Anna about the same time.

After reading Iverson's letter, the Hansen men came to Rock Creek. Twenty-two year-old John, asthmatic and sickly, made the trip in August 1876, and Lawrence arrived nine months later. At Lawrence's urging, John gave up the placer mining he had been engaged in to help prepare for the coming of their women. Lawrence had laid claim to a parcel on Rock Creek, and near that stream, they built a log cabin with a dirt roof and floor, and iron bars, in place of glass, in the windows.

Lawrence and Marie Hansen, with their children, Hedvig and Laura, ca. 1873 *(Courtesy Robert K. Maddock Jr.)*

The women and children —Anna, her sister, Marie, nine-year old Hedvig and seven-year old Laura — came to Idaho in August, 1877. They rode for nine uncomfortable days in a railroad car from Indianapolis to Kelton, Utah, cooking for themselves, and sleeping on wooden seats, cushioned only by the blankets they had brought. Opening windows to relieve the summer heat only brought in clouds of dust, cinders and flies. But, as Anna would later admit, it certainly beat traveling by covered wagon.

John, sun-bronzed and healthy after a year in Idaho, met the women in Kelton with a wagon. After a three-day trip they arrived at Rock Creek which, to Anna, seemed a strange, wild place, beset with dangers, but, as she said, "holding, nevertheless, a glad promise of sweet contentment in a home of her own making."

The Lawrence Hansen cabin

Within a week of her arrival, on September 2, Anna married John in what was likely the first white wedding in what we now call Twin Falls County. Judge William Tranmer performed the ceremony at Lawrence's cabin, with Charles Walgamott and Lettie Dunn—themselves soon to be married—as witnesses. Readers may recall that Walgamott would a few weeks later gun down "Old Dowdle Bill."

The Hansen cabin door had to be removed for use as a table to accommodate the many guests. Since the cabin had only an open fire for cooking, roast beef and other food for a feast was brought from the Stricker store, some three miles away. The wedding cake was baked in Lars Larsen's oven, at his ranch, a mile distant.

That fall, the newlyweds left the homestead of Lawrence and Marie and moved to Oakley Meadows stage station, east of Rock Creek. Here, John taught school to eleven students of various ages. There weren't enough books, and the children had nothing with which to write, except some sheets of wrapping paper and two pencils owned by Hansen.

In the spring of 1878, John and Anna bought a few cows and tried to establish a cheese-making business on some land John had staked out in Cottonwood Canyon, near Rock Creek. But Bannock Indians spoiled these plans. That spring, Chief Buffalo Horn, leading a large war party, burned a stage station and killed at least one man near Glenn's Ferry.

Expecting these marauders to come their way, the Hansens and other families in the Rock Creek area took what possessions they could, turned loose their animals, and fled eastward about forty miles. A few weeks later, after the threat had passed, the Hansens returned to find their cattle gone and the cheese factory destroyed.

Subsequently, John worked about ten years for Herman Stricker as a storekeeper and ranch hand. From 1892 through 1900 he and his family lived in Albion, where he served variously as probate judge, Cassia County superintendent of schools and county

Two Lives of Service

The Hansen brothers had a deep sense of civic duty. Both served for years on the Albion Normal School board, and both were Rock Creek precinct justices of the peace.

John helped create Twin Falls County, and was appointed as one of its first commissioners.

Lawrence was elected to the state legislature and, in 1917, helped obtain funding for the Hansen and Murtaugh bridges.

recorder.

Then, in 1900, he bought the Rock Creek general store from Joseph F. Tatro, who had built it in 1894. Originally, the community known as "Rock Creek" was at the site of the stage station and Stricker store but, in May 1899, or earlier, the post office there had been moved two miles upstream.

Real prosperity came to "Uncle John" Hansen, as he was affectionately known, with the arrival of the engineers and crews who built Milner dam and the canals. His was the only store for miles, and he became a principal supplier to these men. Later he built a store at Hansen, Idaho, the town that bears his name.

Anna and John F. Hansen, 1931
(Courtesy Robert K. Maddock, Jr.)

Range War

—The Sparks-Harrell Empire—

When Col. Frank Hubbs, a foreman for the Sparks-Harrell cattle company, was asked by the Twin Falls *News* how he was getting along with local farmers in the winter of 1905-06, he said he had found them "a square lot of men," who seemed to understand that his company was doing its best to keep its cattle away from their old range.

Just to avoid any unpleasant possibilities, however, Hubbs said he would offer a reward of $2,000 for the arrest of any person who might steal, molest or tamper with any range cattle, no matter where they might be.

Branding in Shoshone Basin *(Courtesy Don Dean)*

Herding Cattle along the Rock Creek Road.

What did he mean by "molesting or tampering?" Hubbs was asked.

"Well," he answered, "if you had seen me trying to pick .22 caliber bullets from the ribs of a thoroughbred Hereford bull the other day you would understand what I mean. Now, that bull was mild mannered an animal as ever ate grass. I have known him for three years, and I never knew him to commit a breach of the peace...I suspect that somebody wanted to use his hide for a gravel sieve."

The previous spring, irrigation water had become available on much of the Twin Falls tract for the first time. But, unfortunately—since no one had told the thousands of cattle and horses grazing in the region that their traditional range was now off limits—they began grazing on the new fields, gardens, and trees.

Because money, materials and time were all in short supply, very little ground had been fenced and, often, the few fences that had been put up were destroyed by the cattle. Consequently, many of the frustrated new farmers had taken to shooting the invaders.

Despite his tough talk, Hubbs could see the handwriting on the wall. Recently he had passed the ruins of old Desert Station, and from there he could see the "pretty" new Twin Falls railway depot. The sight made him realize that, henceforth, the cattle would have to be kept south of the Highline Canal, and that, inevitably, the days of the huge herds were numbered.

Hubbs suggested that someone fence the Highline from end to end, but no one volunteered for the task. So, this little "range war" continued for a few more years, despite the stationing of Sparks-Harrell cowboys along the canal, and the issuance of a county law, which prohibited stock from running at large.

Huge cattle herds had been grazed in southern Idaho since about 1869, coincident with the placer gold excitement on the Snake. As nterest in the placers waned, some of the storekeepers and others moved from the river to establish ranches along the streams on its south side. The Rock Creek-Cottonwood Creek region was settled by smaller, independent ranchers, such as Arthur Norton, Milus Robinson, the Hansen brothers and Robert Brose, to name just a few.

The first big operator to come into our area was Andrew Jasper "Barley" Harrell, a Californian, who, in 1872, trailed some 3,000 cattle over the Sierra Nevada and started them grazing south of the Shoshone Basin. That winter, Harrell's foreman, James Bower, drove

"Honest John"

In 1902, after the death of "Barley" Harrell, John Sparks sold most of his interest in the cattle company to Harrell's son, Andrew, for 1.1 million dollars.

The following year, "Honest John," as he was called, was elected as Nevada's governor, when he ran on the Silver Democrat ticket.

Sparks was re-elected in 1906, but died in office on May 22, 1908. The little railroad town of Sparks, Nevada had earlier been named in his honor.

the herd south onto the lowlands surrounding Wells, Nevada. Later, when Bower realized that the Snake River valley was also good winter range, Harrell brought in more cattle.

John Sparks had been a Confederate cavalryman, fighting Comanches on the west Texas plains during the Civil War. Afterward, he became wealthy raising cattle in Texas and Wyoming. Sparks came to Nevada, and, in 1882, with John Tinnin — who owned the Winecup and the HD brands in Elko County — bought out "Barley" Harrell's holdings in the area.

This huge Sparks-Tinnin outfit maintained a series of pastures, corrals and ranches in southern Idaho and northern Nevada, ranging from Goose Creek on the east to Bruneau on the west, the Snake River on the north and the Nevada desert on the south. Thousands of Herefords and Shorthorns grazed in the hills from early spring until fall, when animals to be marketed were gathered at roundup points, then driven south to the railroad siding at Tecoma, Nevada. In June, 1886 the company bought Tom Canavan's 160-acre homestead along Rock Creek and, on it, established a regional ranch it named the "Shoesole."

In February, 1895, after several severe, cattle-killing winters, John Tinnin sold his share of the Sparks-Tinnin to "Barley" Harrell, and his son, Andrew, and the company became the Sparks-Harrell.

A few months after the change, the Shoesole foreman hired a drifter named Jackson Lee "Diamondfield Jack" Davis, who would soon become Idaho's most notorious cowhand.

Winecup and Shoesole
Branding Irons

A Warm and Fuzzy Cowboy

—The Saga of "Diamondfield Jack"—

Jackson Lee "Diamondfield Jack" Davis was a lovable rascal. Or so thought most citizens of Albion, Idaho, and some of the men who had helped prosecute him.

To Albion residents, Davis was the "assistant jailer." He liked to chat with them through his cell window; he repaired shoes and boys' baseballs, without recompense. He was even known to write bits of poetry for the girls attending the Normal School.

When the convicted murderer's death sentence was commuted to life imprisonment, guards went to Albion to fetch him to the Boise penitentiary. As this news got around town, there was a constant stream of visitors to his cell, most of them bearing flowers, books and other tokens of regard. The deputy sheriff gave Davis a handsome gold watch, and Sheriff Montgomery "Gum" Brown treated him to an elegant dinner. Brown said he had never seen a better inmate than Davis; he was a "great friend of the ladies and an admirer of fine horses."

"Diamondfield Jack" Davis
(Courtesy Idaho State Hist. Society)

But, in truth, Jack was a violent, intimidating braggart of a man.

His was one of the West's most sensational murder cases. It grew out of the bitter fight, between cattle and sheep growers, over grazing rights on the open range. In our area, a "deadline" was agreed upon by both groups, which ran along the ridge separating the Goose Creek and Shoshone basins. The sheep men promised to stay east of the deadline, and the cattlemen to the west. But, in 1895, some sheep herders started crossing the deadline into Shoshone Basin.

That was the year Jack Davis was hired by the Sparks-Harrell's Shoesole ranch at Rock Creek. He had come from Silver City, Idaho, where he likely got his nickname, "Diamondfield," after being taken in by rumors of a "diamond" strike in the nearby foothills, to the extent that he went prospecting for the non-existent gems.

In July, the sawed-off little roughneck was sent to the Shoshone Basin, ostensibly to care for a herd of cattle but, more likely, his job was to keep sheep off the range. In any event, he had several run-ins with herders who had crossed the deadline and, in November, he shot and severely wounded one of them. Jack quickly left the area to prevent the Cassia County sheriff from arresting him on a charge of assault with intent to kill.

Two months later, on February 16, 1896, the gunshot bodies of two young herders from Oakley — John Wilson and Daniel Cummins — were found in their sheep wagon, near Deep Creek, about six miles south of today's little village of Rogerson. The corpses were so badly decomposed that it was impossible to determine exactly when the men had been killed. The date was thought to have been

Attorneys William E. Borah (far left) and James H. Hawley (far right). *(Courtesy Idaho State Hist. Society)*

February 4, the last time they had been seen alive.

It was immediately assumed that Wilson and Cummins had been killed by "Diamondfield Jack," and there was a lot of excited talk amongst the sheepmen of forming a vigilance committee. Within a week, a total of $4,800 in rewards was offered by the county, Idaho's governor and the wool growers' association.

The wool growers hired a young Boise attorney, William E. Borah, to investigate the killings and assist the county prosecutor. The city slicker, Borah — who would later become a renowned U. S. senator — put on some cowboy togs and went out to the Sparks-Harrell ranches. There he learned that two days before the murders, Davis, accompanied by another cowboy, Fred Gleason, had fired recklessly into a sheep camp and killed a horse, not far from the Wilson-Cummins murder scene. And, two days after the murders, Davis and Gleason had been seen carousing in Wells, Nevada, and boasting of having been in Idaho "smoking out sheep herders."

Borah also learned that Davis, alias Frank Woodson, had since been thrown in the territorial prison at Yuma Arizona, convicted of assault with a deadly weapon, and that Gleason was living in Montana under an assumed name.

A full year after the murders, in February 1897, the Cassia sheriff brought both suspects back to Idaho, to face murder charges. John Sparks stood by his men, and, with the help of other cattlemen, hired Boise attorney James H. Hawley—later an Idaho governor—to head the defense team.

Davis' trial was held at Albion in April. The investigation had been bumbled, what with men stomping around the crime scene, and the inept sheriff misplacing much of the evidence. The prosecution's principal witness was a man who had often been in trouble with the law, and who had threatened witnesses and lied under oath.

The evidence against Davis was entirely circumstantial. Borah failed to show that Jack had been present at the crime scene on the assumed day of the killings. Davis and Gleason had been seen leaving the Brown Ranch on Salmon Falls Creek at sunrise, and they arrived at the Boar's Nest Ranch, near today's Jackpot, Nevada, about 1

Jeff Gray (seated). The man standing is believed to be Shoesole foreman Charles Hewitt. *(Courtesy Lucy Gray)*

P M. Hawley argued that they could not have ridden from one ranch to the other, via the murder scene—some 45 miles—in so short a time.

Despite the flawed case, Jack was found guilty and was sentenced to be hung. In a separate trial, a week later, Fred Gleason was found not guilty of the same charges. It thus appeared that Davis had been convicted simply because of his animus toward the sheep men.

Davis' legal battle would continue for another five and a half years. His lawyer made three appeals to the Idaho Supreme Court, and one to the U. S. Supreme Court. Four times, the state heard applications for a pardon or commutation of sentence. Jack's execution was stayed eight times.

Not long after Hawley's first appeal to the Idaho Supreme Court, he was able to confirm his suspicion that James E. Bower — a former Sparks-Harrell superintendent and a respected citizen — had been involved in the murders. Hawley brought pressure to bear on John Sparks to get the truth out, and soon thereafter Bower revealed to Hawley that Jefferson D. Gray, of Rock Creek, had killed Wilson and Cummins. Bower claimed it had been done, in self- defense, during a confrontation and struggle Bower and Gray had had with the two sheep herders over range rights.

Both Bowers and Gray swore affidavits confirming this, and both were charged with the murders. The cattlemen paid Hawley to defend them and, in perfunctory trials, both were acquitted.

The revelations of Bower and Gray had no effect on Jack Davis' status, even though they were confirmed by other evidence. Many people, including Borah and the pardon board, thought that Bower and Gray's affidavits were simply a scheme to save Davis' life.

In July of 1901, however, Davis' sentence was commuted to life in prison, and he was sent to the state penitentiary. On December 17, 1902 he was given a full pardon. That afternoon Jack, the warden, and lawyer Hawley left for the statehouse, in a horse-drawn cab, for the pardoning ceremony. As they approached Boise's fine watering place, the Natatorium, a stop was suggested, and the three went in for several rounds of drinks before proceeding to the capitol.

Done in by a Taxi

After his pardon and release from prison, Jack Davis made a fortune in the Tonopah, Nevada mining district, starting with a grubstake from one of his former prosecutors.

There, he was involved in several anti-labor union fights, frequently flourishing knives and guns to threaten his opponents. Later, Davis lost most of his money and, in December 1949, he was struck down and killed by a taxicab in Las Vegas.

The Railroad Comes to "Big Bottom"

—The Beginning of Tourism at Shoshone Falls—

One woman who saw Shoshone, Idaho frequently during its early years—Carrie Adell Strahorn—called it a "disappointed railroad centre." Her husband's work had taken them to a lot of rough new towns, but she had seen none so bad as Shoshone, with its bawdy dance halls on every corner and gunfire at all hours. And, the only place to put the many lawbreakers was a hole in the ground, with guards posted around it.

Charles Walgamott

Another visitor called the region around Shoshone so barren that the cows eyed "every tomato can with greedy interest." He complained that passenger trains passed in both directions "only in the dead of night" — just as they did, in our time, before Amtrak service was discontinued.

Officials of the mighty Union Pacific Railroad were spurred to extend their line eastward across sparsely-populated Idaho Territory when an Oregon company announced plans to do so from the west. As a preemptive action, the Union Pacific quickly incorporated a subsidiary, the Oregon Short Line, to build and run such a railroad.

Construction of the Short Line began in July, 1881 at Granger, Wyoming, with crews driving steadily westward, conforming, generally, to the Oregon Trail. In February, 1883 the tracks reached a brand-new Idaho town known, variously, as "Big Bottom," "Naples" or "Junction City." Track continued to be laid westward toward Oregon, and soon the Short Line would carry most of the freight and traffic into southern Idaho, making the wagon road from Kelton, Utah obsolete.

Walgamott and Sullaway's Tent Hotel at Shoshone Falls
(Courtesy Idaho State Historical Society Library)

By that autumn, "Big Bottom" had become Shoshone, and was growing apace. The railroad had built a depot and roundhouse, and was planning to erect car repair shops. There were about 150 "houses" of canvas, adobe and rock, strung along the tracks. A branch line was being built northward to service the new Wood River mining towns of Bellevue, Hailey and Ketchum.

Shoshone's boosters were certain that railroad advertising would soon bring throngs of tourists to see Shoshone Falls, and that, within five years, their new town would become "the Denver of Idaho."

By the time of the railroad's arrival, young Charles S. Walgamott — he of "Old Dowdle Bill" notoriety — was in possession of the land surrounding the falls. Seven years earlier he had established a squatter's right to 160 acres on the south side. Recently, he had formalized that claim, and had filed on another parcel, and some spring water, on the north bank.

He and a partner — a stage-coach driver named Joseph Sullaway — made a passable road from the northern rim rock down to the river and, that summer, began running an eleven-passenger coach daily from Shoshone. They also established a hotel of tents, charging $1 per night. They gave customers the option of sleeping on the ground or paying an extra dollar for a hammock or cot. Those tourists reluctant to spend the dollar were told bedtime stories by Joe Sullaway, about how rattlesnakes often crawled into people's blankets.

In September of 1883, Walgamott sold his property on both sides of the river to Emerson Stone, a wealthy furniture dealer of Omaha, Nebraska. Apparently Stone agreed to let him to continue to run the tent hotel. In April of the next year, Walgamott and the "Shoshone Falls Company" — probably a partnership with Stone — were granted a license by Alturas County to operate and maintain a ferry at the falls.

Shoshone Falls Ferry, about 1910

A Scary Ferry

At first, Walgamott's ferry consisted of a rowboat with a safety rope strung across the river. The crossing was only 300 yards above the falls, where the "wondrously strong" current always caused concern about the boat breaking loose.

Only two passengers could be taken across at a time, and Walgamott always made them look at him, not the water, until the boat had crossed the main stem of the current, lest they panic and cause the boat to founder.

At the time Walgamott established the ferry it was the only one for about 30 miles in either direction, but tourists and visitors were scarce, and Walgamott spent a lot of time "praying for somebody to come." Among the few that did was a man destined to have a huge impact on the future of southern Idaho.

According to Walgamott, one night in early October, 1884, a young fellow named Ira Burton Perrine arrived at the tent hotel driving a small herd of cows he had purchased in Pocatello, intending to go into the dairy business near Hailey. Perrine was looking for winter range for the animals. Walgamott invited Perrine to stay the night, and in the morning told him about a place a few miles downstream, where the canyon had ample pasture, as well as two lakes that never froze up.

These, of course, were the Blue Lakes, where Perrine would soon begin to carve out his future.

A "Charming and Attractive Resort"

—Burt Perrine's Blue Lakes Ranch—

In the fall of 1883, there arrived in Idaho a young tiger from Indiana named Ira Burton Perrine. Twenty-two year old "Burt" was eldest of the eleven children of George and Sarah Burton Perrine, which may explain why, after a year of college, he came west to live with an uncle and aunt at Bullion, a booming new mining camp near Hailey.

Blue Lakes Ranch in 1905

Weighing a mere 135 pounds, soaking wet, and standing only five feet four inches high, the young go-getter quickly set about molding Idaho Territory to his liking.

It is uncertain how Perrine first learned of the beautiful Blue Lakes property, but he probably found it while looking for placer mining locations. As we have seen, Charles Walgamott took credit for pointing out the place as pasture for Perrine's dairy cows.

But, if indeed Perrine owned a dairy, and if Walgamott told him about the Blue Lakes, it wasn't in October 1884, as claimed, because six months earlier Perrine and a friend, Thomas Hyndman, had filed on all the water in the Blue Lakes, for use in mining, irrigation, and fish culture. Also that spring and summer, Perrine was mining in the Hagerman Valley with another partner.

In any case, find the Blue Lakes he did. Since, at first, the only way to get down into the canyon was by a narrow foot trail, Perrine and Hyndman used a rope windlass to lower lumber and other materials for the building of a small shack in which to live.

The stunningly crystal-clear Blue Lakes sit well above the level of the Snake River. The upper of the two is fed by springs, and flows into the lower lake. From there, the water sinks to run underground for about a quarter of a mile, and then bursts forth from several places to form a large, sparkling stream that rushes along a rocky channel for over a mile before emptying into the river.

Perrine and Hyndman dug ditches to carry water from the lakes to the orchards they began to plant. Initially, it seems, Hyndman did most of the work of developing the ranch, while Perrine

Stagecoach going down to Blue Lakes Ranch

sought the money to do it. He may have worked several winters in the mines at Butte, Montana but, by 1886, was living in Shoshone, running a livery stable and commuting to the Blue Lakes as needed. That year he filed on Little Wood River water, above Shoshone, and may have done some mining and farming on the south side of that stream.

By 1889, Perrine and Hyndman had the Blue Lakes property pretty well tamed. Over 3,000 fruit trees had been planted, as well as grapes and other small fruits. They had built a usable wagon road out of the canyon, and had stocked trout in the lakes and in the huge stream that gushed out of the canyon wall. They had begun marketing fruit in Shoshone, Oakley and the Wood River valley.

So, that summer, Perrine went into the tourist business. Three years earlier, a decent hotel had been built at Shoshone Falls and, thanks to railroad advertising and special fares, it was drawing quite a few visitors. Perrine was determined to cash in on its popularity, by making the Blue Lakes part of any visit to the Falls.

He bought an old stagecoach that had once run on the Kelton-Boise line, and began taking tourists — many of them Boiseans — on excursions from Shoshone to Hailey, Shoshone Falls and the lakes. In 1891, he prohibited camping on his property and opened a small hotel there, adorning its walls with oil paintings ordered from Oregon.

Business was given a boost that summer by the visit of the first of many celebrities Perrine would come to know. The "Little Wizard" — financier Jay Gould — was vacationing with his family in the Wood River country, his private railroad car parked alongside Silver Creek. In late August, while visiting Shoshone Falls, Gould was told about Perrine's "charming and attractive resort" which, at the time, was getting quite a few visitors.

The magnate made a surprise overnight visit to the lakes, and came away delighted by the wonderful fishing and the delicious dinner of trout and Blue Lakes fruit. One of his party christened the stream that flowed just in front of the hotel, "Gould Creek," in honor of the distinguished guest. Perrine assured Gould that the stream would always retain that name, but later it was called Trout Creek, and today is known as Alpheus Creek, after a mythological river god who pursued the nymph Arethusa through the bowels of the earth.

Blue Lakes Water

In early Twin Falls — when potable water came either from a shallow well in Rock Creek canyon, or from the lowline canal — the brothers Louis and Frank Benoit began selling "Blue Lakes Spring Water." It was brought to town in a large tank and sold by the gallon. Later, the Benoits bottled carbonated water and soft drinks.

As early as 1905 there was talk of pumping Blue Lakes water across the river for use in Twin Falls, but not until 1967 did this begin. Today, a substantial amount of the city's requirements come from this source.

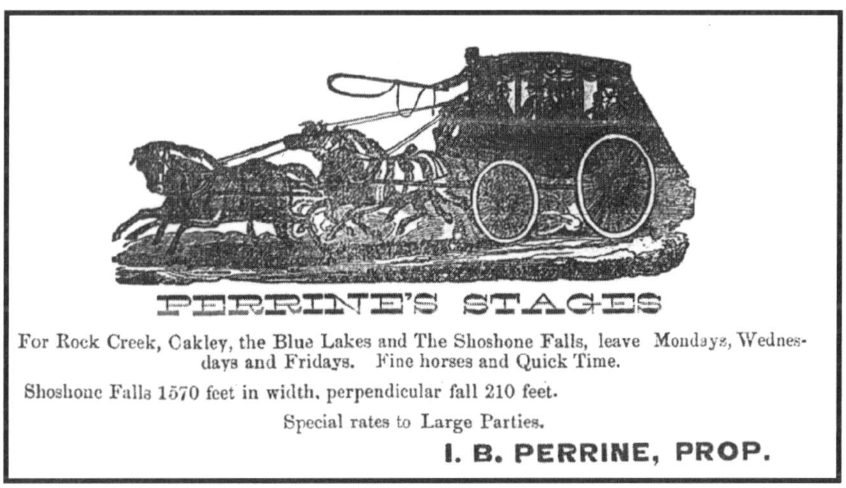

Advertisement from *Shoshone Journal*, April 21, 1899

The Farmer Takes a Wife

—The Perrine Family—

At the Blue Lakes Country Club, not far removed from a busy golf cart path, sits an obscure little cemetery. Within the boundary of its tilted old iron fence lie the remains Ira Burton Perrine, his faithful and devoted wife, and one of their children. Husband and wife now rest where they once labored so hard to reclaim the land from nature's desert.

By 1892, Burt Perrine was running the Blue Lakes farm and hotel, his Blue Lakes and Shoshone Falls Stage Line, and an ice house in Shoshone. He was also still heavily involved in placer mining ventures, and he and a fellow named William Keyser were owners of the Lincoln Hotel in Shoshone, which they had purchased in April from Donald MacKay.

So busy was Perrine that it is hard to imagine just how he found the time to woo and win any woman's hand. But, in June, he married Donald McKay's seventeen-year old daughter, Hortense Amanda, who had just graduated from an Episcopal girl's school in Salt Lake City. Perhaps what won her over was the naming of one of his Augur Falls placer claims "Hortense," in her honor.

Hortense Perrine and daughter Stella, ca. 1900.
(Courtesy Twin Falls County Historical Museum)

The newlyweds were properly chivareed by their friends and, after a brief honeymoon in Salt Lake City, they went to live full time at Blue Lakes. In later years, Mrs. Perrine always liked to tell the story of how, at sunset on a beautiful June evening, her husband took her down into the canyon to become mistress of the ranch.

The first of the four children of the union, George Burton Perrine, was born the following year, and a daughter, Stella, arrived in November, 1895. A few months before Stella's birth, Perrine bought out the interest of his Blue Lakes partner, Tom Hyndman.

Perrine continued to plant fruit trees and develop the gardens and fields of his canyon property. He marketed loads of his fruit in Challis, as well as in Hailey and Shoshone, where he owned grocery stores.

He seized upon every opportunity to publicize the Blue Lakes in order to drum up business for the hotel and stage line, often hosting newspaper editors and other men of note. Fruit was sent to the various expositions that were so common in those days, and the many awards Perrine won added

Springtime at Blue Lakes

In 1905 Hortense Perrine began holding annual "peach blossom," parties at Blue Lakes.

The first of these gatherings included only Mrs. Perrine's closest lady friends but, later, many more were invited to come. The event soon became one of the social highlights of the year.

The parties were held at the peak of the blossoming season when a number of different kinds of fruit trees were in bloom at one time.

The ladies were usually given a carriage ride to the lakes, where they could pick wild flowers and watch the fish frolic in the clear waters.

Then they came down into the orchard to inhale "their fill of the sweet scented atmosphere" while dining on food cooked over a campfire.

to the reputation of his enterprises.

Drawn, perhaps, by Burt's growing prosperity, his younger siblings began migrating to Idaho. The first to come, sometime in the early nineties, was Silas D. Perrine. By 1896, Silas had either purchased, or was managing the stage line, which was by then running four and six-horse coaches tri-weekly from Shoshone to Rock Creek, Oakley, Blue Lakes, Shoshone Falls.

To open a new route for the stages to Rock Creek, in 1897 Burt began making a wagon road out of the canyon on the south side, opposite the Blue Lakes. Two years later, when it was finished, he put in a ferry on the river.

Another brother, J. Frank Perrine, arrived from Chicago in April 1896 and stayed alternately in Shoshone with Silas, and at Blue Lakes with Burt. He drove one of the Perrine stagecoaches and, in June, 1897, reportedly bought the line from Silas, who began managing a ranch near Shoshone.

Emma Perrine, the frail, 27-year old unmarried sister of Burt and Silas came from Chicago to live at the Blue Lakes in the fall of 1898, hoping to regain her health. The following year, a cousin, Silas D. Burton, arrived from Anaconda, Montana, with his wife, and began working for Ira.

But, the closely-knit Perrine clan was seriously decimated, when death took three of its members within an eight-month period, near the end of the decade. Burt and Hortense's five-year old, George, fell into an irrigation ditch while his parents were picking peaches in their upper orchard. He became ill and was diagnosed as having "membranous croup" (bacterial tracheitis), and died a painful death on September 16, 1898. The funeral was held at home, after which they buried little George not far from the place where his brief time on earth had been spent.

The surviving Perrine children in 1912 when they were called back to Indiana by the death of their father. Front row: Charles, Lena, May, Ira. Back row: Silas, William, George, Ledyard, and Walter.

Just a few months later, in December, 22-year old Frank died at brother Silas' home five miles west of Shoshone, reportedly of "brain fever" resulting from a severe cold.

Blue Lakes Cherry Trees in Bloom.

Meanwhile, Emma had contracted influenza, which failed to respond to treatment. She went to Shoshone to stay with Silas, in May 1899, where she died suddenly of a pulmonary hemorrhage. Both rank and Emma were taken home to Indiana for burial.

Despite the heartbreak of these losses, Burt Perrine continued to work incessantly, undertaking one new project after another. And, over the next few years, still more of his kinsmen would join him in Idaho, to share in the fruits of that greatest of his enterprises, the reclamation of a quarter million acres of desert.

Ira Burton Perrine

Vacationing At "Little Dewey House"

The Little Dewey House, ca. 1906.

Visitors to Shoshone Falls during the "Gilded Age" of the 1890s could view the grand spectacle from a rocky ledge called "Lover's Leap." The highly-inventive promoters for the Oregon Short Line Railroad told prospective travelers that, "once upon a time," an Indian girl and her lover had leapt to their deaths from the brink of the cataract. It was said that "Now...the maiden comes when the moon is bright, and crouches upon the rock, the breeze playing with her long, black hair."

After buying Charles Walgamott's property at Shoshone Falls, in 1883, Emerson Stone had acquired more land there, on both sides of the river, and then sold an interest in the property to three others — his partner in the furniture business, Charles H. Dewey; an Omaha banker named John Creighton; and William A. Clark of Montana — all millionaires.

GRAND OPENING OF THE NEW HOTEL!

AT

SHOSHONE FALLS, IDAHO!

JULY 15th, 1886.

THE UNION PACIFIC RAILWAY

Announce the following Extremely Low Round-Trip Rates to the Falls and Return, for this occasion:

From		From	
Salt Lake,	$17.15	Butte,	$23.75
Ogden,	15.80	Anaconda,	24.40
Park City,	18.00	Bellevue,	6.80
Evanston,	18.35	Hailey,	7.10
Logan,	13.90	Ketchum,	7.75
Pocatello,	9.75	Caldwell,	11.80
Soda Springs,	14.35	Boise City,	14.80
Blackfoot,	11.70	Huntington,	15.70
Eagle Rock,	12.40	Baker City,	18.20
Dillon,	20.00		

Tickets Good Going July 12, 13, 14 and 15; Good Returning until July 22.

All Tickets from points north and south of Pocatello include a side trip to Soda Springs, and return, without extra charge. Holders of these tickets desiring to include the Soda Springs Trip, will present their tickets for proper indorsement to the General Agent of the Ticket Department, who will be at the Hotel Opening on the 15th.

Railroad rate schedule for opening of Little Dewey House. *(Author's Collection)*

In 1886 these men built two hotels — the "Dewey House" in Shoshone, and the "Little Dewey House," on the south side of the river at the falls. The two-story "Little Dewey" sat on the bluff, near the present overlook, and contained 23 "spacious rooms, well furnished with all modern conveniences." Sleeping there, with the roar of the falls, and the trembling of the building was said to be a unique experience.

Typically, visitors coming by rail would stay overnight in Shoshone, and take a stagecoach the next day to the river. The railroad's promotional materials even tried to romanticize this dreary 25-mile trip across the desert, which offered nothing to see but jackrabbits and an occasional horseman.

The trip, it was said, was made in an old Concord stagecoach, with a "faded plush interior," and was conducted by "one of those thoroughly Western characters, a Rocky

Mountain stage driver." The tall, angular driver was "sunburned and jovial, full of stories and border vernacular," and was known to "take a drink when pressed, and loftily smoke a proffered cigar, holding [it] at a ferocious angle and eating about half of it."

Charles Haynes driving passengers down into the canyon.

The driver described was probably Charles C. Haynes, a competitor of Ira Perrine. Haynes had been in Shoshone since 1889 running a stage service, intermittently, to the Falls, Blue Lakes and Rock Creek The old "Jehu" – so called after the Old Testament character who did "driveth furiously" – had been piloting stages since he was 18 years old, and had worked on a multitude of lines – the Kansas Stage Company, the Overland Mail, Wells-Fargo, and the Northwestern Stage Company on the Boise-Kelton road. He had also run lines of his own from Battle Mountain to Tuscarora, Nevada, and from Ketchum to Sawtooth City.

Arriving at the river from Shoshone, stages would take the rough road down the north wall of the canyon to a short stretch of sand and the ferry landing. The stage was carried across to the south bank, where, a few rods from the landing, was the Little Dewey House. One could stay there for a dollar per day, with children at half that price.

In addition to "Lover's Leap," the railroad's advertisements contained fanciful names for other features near the falls. There was a "Wonderful Trail," the "Devil's Spade" and a "Profile Rock." At the foot of the falls one could visit "Diana's Bath"—a natural pool within a large cave, fed by cold spring water dripping from the grotto's high dome. The

Over The Brink

A number of persons have been swept to their death over Shoshone Falls, but one of worst episodes occurred on March 30, 1904, when three people perished. That evening, two men rang the call bell on the north side for the ferry to come and get them. Marie Willis, a waitress at the hotel who occasionally ran the ferry, took it across to get the passengers. Once there she found that the draft rope was out of order.

Another employee, Sam Graham, and an unidentified man, left the hotel and crossed the river in a small skiff to lend assistance. Unable to fix the problem that night, Graham, Willis and the unnamed man started back across the river in the skiff. With three passengers the little boat sat very low in the water. Added to this, the river was rough from brisk winds. When high waves began swamping the boat, its occupants panicked and, capsized it.

Cries for help were heard as all three people were carried down and over the falls to their death. Despite an extensive search, including the use of drag hooks and dynamite, the bodies were never found.

ads failed to mention that to get to the cave, one had to descend from the rim on a steep, rocky trail, hanging onto a grab rope for dear life.

At the time, the bluff upon which the hotel stood was overgrown with junipers. There were pathways through this grove to Lover's Leap and the other points of observation — Prospect Point, immediately in front of the hotel site, and Lookout Point about 200 yards down river. Elsewhere there was a Point McKinley, Point Garfield, Dewey's Bluff and a Natural Bridge.

Just upstream of where the river makes its last grand leap, are a number of lesser falls and cascades that, in those days, bore names like Bridal Veil, Bride's Maid, Two Graces, and the Mill Race. Among the rocks and islets interspersed amongst these cascades is Bell's Island — so named for Tom Bell, a placer miner who, in 1881, was carried over the falls, along with three Chinese men he was attempting to ferry across the river in his rowboat.

Wildlife abounded in the canyon in those days. One of the unique features above the falls is a pillar called Eagle Rock, which rises from the rapids about fifty yards from the south bank. For many years, pairs of bald eagles could be seen there, tending a nest made of white, water worn sticks. And, amazingly, for a time in the summer of 1896, two large sea lions were seen frolicking in the river and taking the sun on the rocks just below the falls.

Shoshone Falls and hotel, ca. 1907

The Great Commoner

—Burt Perrine and Williams Jennings Bryan—

In our county historical museum there hangs a piece of rough wood with this little verse inscribed on it:

Shoshone Falls, a work divine
Present an awe-inspiring view,
And teach that God is great.
But, from our trip to Ranch Perrine,
We learn what human hands can do.
Would we could imitate.

Bryan gazes wistfully at the White House in 1899 political cartoon. *(Author's Collection)*

The words are those of William Jennings Bryan, written in the year 1897, when he and Ira Burton Perrine began a warm friendship.

A year earlier, the famed Nebraska Democrat had been chosen as his party's presidential candidate. At the time, he was a leading spokesman for the unlimited coinage of silver at a ratio to gold of 16 to one, and it was his famous "Cross of Gold" speech, made at the party's convention, that won him the nomination.

The handsome, vigorous Bryan was surely the most compelling orator of his time, and, when he spoke at the convention, his marvelous, baritone voice soared, unamplified, to the farthest recesses of the assembly hall.

In extolling free silver coinage he spoke disdainfully of "organized wealth," and of the hardships on debtors that a gold standard would cause. He concluded his oration with a promise that working men would answer the demand for a gold standard by saying to the wealthy — "You shall not press down upon the brow of labor this crown of thorns; you shall not crucify mankind upon a cross of gold."

"Boy Bryan" thus became one of the youngest men ever to run for the presidency. In the ensuing campaign, he worked tirelessly, traveling through 27 states, and speaking up to twenty-five times a day. But, despite this effort, William McKinley won the election, decisively.

The following summer, Bryan presided over a "Trans-Mississippi Congress" at Salt Lake City. It was there that Idaho U. S. Senator Fred Dubois (a Silver Republican who later turned Democrat) introduced him to Ira Perrine. An invitation to visit the Blue Lakes was extended and, a few days later, on July 24, 1897, Bryan and his family arrived at Shoshone on the morning train.

The "Great Commoner," as he was known, had received 78 percent of Idahoans' votes for President, so his arrival drew a large crowd of local families, as well as many from the Wood River country, who had come down on the branch railroad the previous evening.

Bryan and "Buffalo Bill" Cody argue over who is the better showman. *(Author's collection)*

The guests were served breakfast at Matt McFall's New Columbia Hotel, and then went to the Shoshone Pavilion where, though hoarse from making many speeches, Bryan addressed the crowd. Afterward, Perrine escorted his visitors across the desolate plain to the Falls, and thence to the Blue Lakes, in a six-horse coach driven by "Captain" Charles Haynes, a former stage driver for the Overland Mail Company.

Perrine and his guest hit it off right from the start. They were both the same age (37), from the Midwest, and were hard-working and optimistic men.

No doubt Bryan was fed plenty of fried trout and fresh fruit, and he may have been taken on a moonlight visit to Perrine's melon patch. He probably was rowed up river to see the "Natural Piers" (Pillar Falls), and the eagle nests so common in the canyon those days.

In any event, Bryan was totally delighted and refreshed by the visit. He remarked that Perrine's ranch —"as green as a Nebraska farm in June"— demonstrated the possibilities of irrigation, and testified to Perrine's intelligence and industry. The views were entrancing — the great spring issuing from the side of the hill, the stream rushing down through a series of falls to the river and, prettiest of all, the Blue Lakes — so clear that one could see the trout rise from the rocky bottom to take a fisherman's bait.

The "Monkey" Trial

In July 1925, the deeply-religious Bryan helped prosecute John Scopes, a Tennessee teacher who had purposely broken that state's law prohibiting the teaching of Darwin's theory of evolution in public schools

This "monkey" trial attracted worldwide attention, the press portraying Bryan as a "pitiable punch drunk warrior" who was clearly no intellectual match for the famous defense counsel, Clarence Darrow.

Scopes was found guilty, at the request of Darrow, who wanted the case to be heard in the state supreme court. That body later reversed the conviction on a legal technicality.

Five days after the trial ended, Bryan died quietly in his sleep, exhausted by his busy life, and weakened by diabetes and a recent automobile accident.

He lies buried in Arlington National Cemetery beneath the inscription, "He Kept the Faith."

In 1900, after Bryan once again lost the presidential election to McKinley, he founded and edited a weekly newspaper, *The Commoner*. In it, he began championing a wide range of progressive reforms — workman's compensation, an eight-hour work day, government regulation of railroads and utilities, and women's suffrage, just to name a few.

On September 14, 1907, nearly ten years after his first visit, Bryan and his wife came once again to the Blue Lakes, as guests of the Perrines. By then Twin Falls had sprung into being, and had its own railroad and depot, where the Bryans were warmly received by a large delegation and the city band. On his drive to the canyon Bryan was amazed to see the prosperous new town and the farms, sitting in the midst of what had previously been only rangeland.

As before, Bryan enjoyed his stay immensely. He fished a lot, and sat in the orchards working on a book he was writing. He explored the area, and built three rock monuments on a cliff overlooking the Perrine home. He also visited the new north side tract, which was about to be thrown open to entry, and he arranged to buy forty acres adjoining the Jerome town site.

On the final day of his visit, September 20, before catching the evening train, Bryan delivered a speech from the veranda of the Hotel Perrine. The weather was beautiful, and every square foot of the spacious hotel court was occupied, as were the surrounding streets.

Bryan was introduced by Senator Dubois — by then a card-carrying Democrat. In a lengthy speech, Bryan praised Perrine for helping turn the desert into a garden, and making homes for thousands of people. Wealth drawn from Mother Earth, he said, was legitimate wealth, unlike that gained by exploitation of the public.

The following year, 1908, Bryan ran once again for President, and was once again defeated, this time by William Taft. He thus became the only man ever to run for the office three times without winning.

Bryan speaking at Hotel Perrine in 1907. Ira Perrine is seated on Bryan's immediate left; then Robert McCollum, a business associate of Perrine, and George Fraser of the *T. F. News.* Fraser, a Republican, seems to have lost interest in the speech.

Whose Idea Was It, Anyway?

—Twin Falls Irrigation Project Conceived—

Within a few years of the completion of Milner Dam in 1905, and the shunting of Snake River water into canal systems, the desert plain bloomed, as the rich, fertile soil was grubbed free of sagebrush and made arable. Within a few years, our region was being called "Magic Valley," because of the wondrous, transmuting effect of the water.

The Cedars area, just downstream of where Milner Dam would be built. Clappine Rock is in the center. Note the waterwheel, used for placer mining and irrigation.

Ira Burton Perrine is usually credited as being first to realize that "The Cedars" — as Milner was then known — was a natural place to divert water to irrigate a vast tract of land on both sides of the river. One source has the idea dawning on him while camped there on one of his many trips taking fruit to sell at Albion. "While thus camped one night with a sagebrush fire and a Dutch oven for a foreground, the little man saw in the flames the future..." so goes this story.

This is romantic stuff, but, in fact, the possibility of using The Cedars was probably first noted by someone else. The location is a natural one for a diversion dam, being, as it is, just above the place where our long, deep canyon begins. Also, the river there was divided by two large islands and is narrow enough that, in1890, the people of the region tried, unsuccessfully, to raise money for a bridge at that point.

Some historians believe that Clappine Rock, just below where the dam would come to be built, marks the beginning of the treacherous "Caldron Linn," where Antoine Clappine, a member of the Wilson Hunt expedition of 1811, was drowned.

Arthur D. Foote, a prominent civil engineer, could have been the first person to have noted the potential at The Cedars. He supervised the Idaho portion of a huge survey conducted in 1889 and 1890 by the U. S. Geological Survey, under John Wesley Powell, whose purpose was to find arid lands suitable for irrigation, and to locate likely places for storage reservoirs and canals.

Most of the Snake River reservoir sites and canal lines located during the 1889-90 surveys were in eastern Idaho. If Foote did realize the potential of The Cedars, he made no mention of it in his official reports. In fact, he said, at the time, that the river below American Falls ran continuously in a canyon

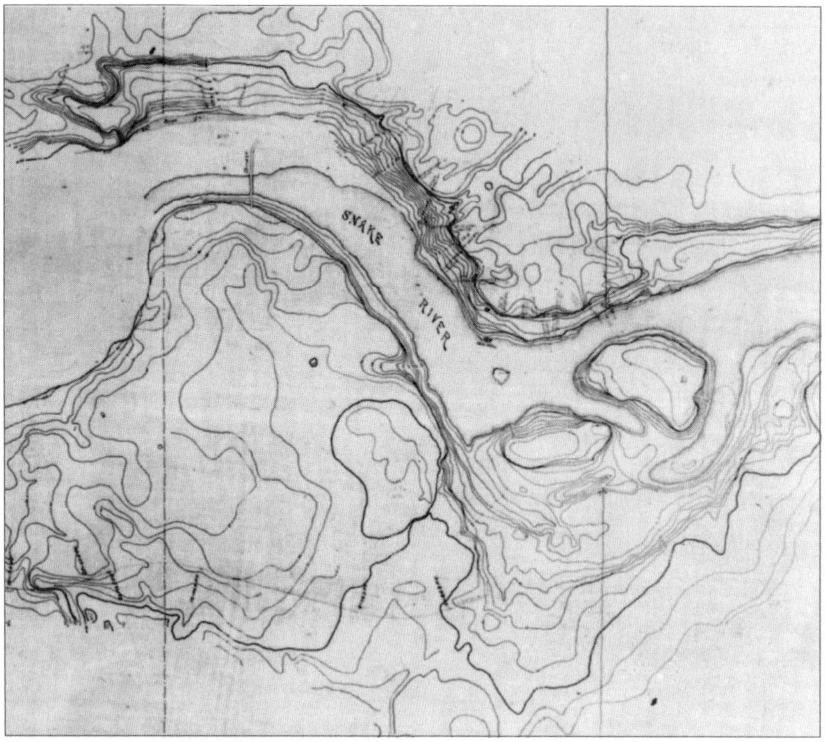

Topographical map of The Cedars section of Snake River.
(Author's collection)

too deep to be tapped for irrigating any part of the counties now known as Twin Falls, Cassia, Owyhee, Elmore, Lincoln, Jerome and Minidoka.

Nonetheless, D. W. Ross — Idaho's state engineer during the planning for the Twin Falls irrigation project — said Foote spoke of having noticed that a canal could be taken out near The Cedars, but that he, Foote, had never gotten an opportunity to test his theory.

An early editor of the *Twin Falls News* also gave Foote the credit, saying that William J. McConnell —who later became Idaho's governor — and others who had faith in Foote's judgement, formed a company and spent $20,000 making plans for an irrigation system. The financial panic of 1890 supposedly compelled them to abandon the undertaking. The editor also said that Ira Perrine became familiar with Foote's surveys.

It may be that Perrine was even a part of McConnell's project. In April, 1890 the *Shoshone Journal* reported that an Eastern syndicate was about to close a deal to "handle one of the largest irrigation enterprises in Idaho." It was also alleged that the "Shoshone Falls Power and Electric Railway Company," had just been formed to develop Shoshone Falls' immense water power. The town of Shoshone would be lighted, and an electric railway would be built to carry tourists to the falls.

No names were given, but these schemes were probably McConnell's. The article's place of origin and the mention of an electric railway — a lifetime obsession with Perrine — hint of his involvement.

Another likely person to have first seen the possibilities at The Cedars was Frank Riblett, Cassia County's first elected surveyor who, early on, had a vigorous interest in irrigating the region. Beginning in 1883, he, George and T. A. Starrh, and others, began claiming enormous amounts of Snake River water for various projects. Their earliest scheme would have built a dam at the very place where Minidoka Dam would later be built by the federal government. It evidently came to naught because financing couldn't be found.

At The Cedars, Riblett attempted, unsuccessfully, to start an

Mary Hallock Foote

Arthur Foote's wife, Mary, was a talented illustrator and author. She wrote four novels and nine short stories, based on her experiences living in the Boise area. Eleven of her wood engravings, called "Pictures of the Far West," appeared in *Century Magazine.*

In her reminiscences, Mary wrote that her husband camped at The Cedars, and saw how the levels ran, and predicted that the Snake would, before long, be diverted at that point to irrigate the desert.

irrigation project a full five years before Ira Perrine began formulating such a venture. In the spring of 1895, he filed on water at that exact spot, to be used for irrigation in both Lincoln and Cassia counties. Riblett intended building a dam across the river "at the head of the falls known as Geyser or Cedar Rapids," about a mile below the old stagecoach station called the Stone House. His canal would have run southwesterly, just as does the one that Perrine — a man of action, not a dreamer by a campfire — eventually got built.

Milner Dam as it looks today.

Formation of The Twin Falls Land and Water Company

—Irrigation Project Nixes Plan for Shoshone Falls National Park—

Ira Burton Perrine died in 1943, at age 82, after suffering a stroke a few weeks earlier. According to his biographer, while hovering near death in his hospital bed he overheard a nurse comment that the room was a little chilly. Seeing his lips moving, one of those present leaned in close and heard him whisper, "You should pipe the hot water at Yellowstone Park and bring it here. Then you would always be warm."

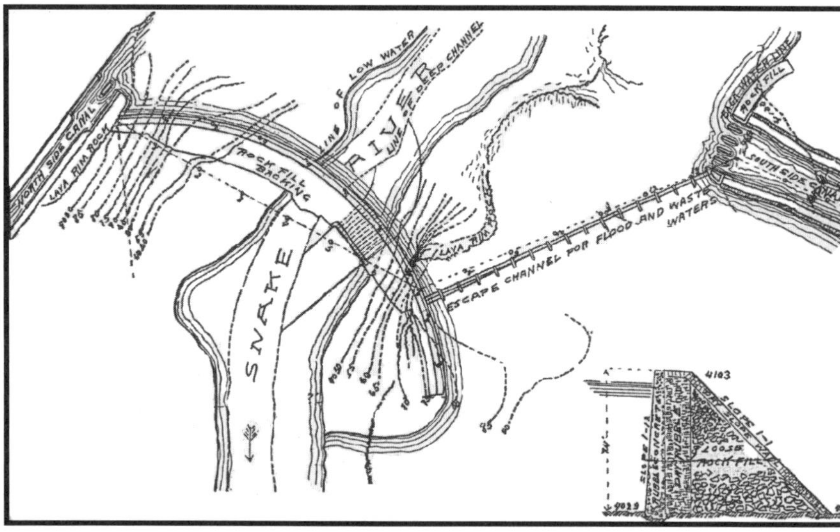

First conceptual sketch of Milner Dam.
(From Boise *Statesman*, Jan. 1, 1902)

While this story may be fanciful, the spirit it discloses was very real. The dying old man, forty-three years earlier, had been in full stride, trying to start a project equally as complex as piping hot water from Yellowstone.

The Carey Act of 1894

The Carey Land Act, named for Senator Joseph M. Carey of Wyoming, provided for turning over certain federally-owned desert lands to Western states upon the condition that they be developed for irrigation.

The states were expected to make contracts for construction of the irrigation works with private parties, who would be repaid for their work by the sale of water rights to the settlers.

Any U. S. citizen, over age 21, except married women, could acquire up to 160 acres of Carey Act land. The state sold an entryman the land, and the builder of the irrigation system sold a corresponding number of water shares, all for a nominal cost and with generous payment terms.

It isn't known exactly when Burt Perrine started promoting — or revisiting, as the case may be — the idea of taking water out of the Snake River at The Cedars (Milner). It could have been as early as the summer of 1899, when Carey Act projects began to proliferate in Idaho.

He turned first to Stanley B. Milner, of Salt Lake City—his partner in some failed placer mining ventures a few years earlier—for help in organizing and financing the irrigation project.

Sometime after disclosing the idea to Milner, Perrine approached the state engineer, D. W. Ross, on the subject. Ross went to The Cedars with Perrine for about a week in early June 1900, and came away convinced that diversion of water there was, indeed, feasible.

Gov. Frank Steunenberg
(Courtesy Idaho State Hist. Society)

Ross also may have made an agreement with Perrine that each of them would obtain half of the required money from his circle of friends.

Soon thereafter, Perrine staked out two water claims at The Cedars, one on either side of the Snake, for purposes of irrigation and mining. Each claim was for 3,000 cubic feet per second (c.f.s.) of flow. The filings were sent to Ross, who, revised them downward to allow diversion of 3,000 c.f.s. on the south side and 400 c.f.s on the north.

Ross called the project to the attention of Idaho's governor, Frank Steunenberg, who agreed to help obtain the necessary capital. Subsequently, James H. Lowell and the governor's brother, Albert K. Steunenberg, were told about the scheme. Lowell was a well-known civil engineer who had worked on many irrigation ventures in the Boise area. The Steunenbergs were publishers of the *Caldwell Tribune* and owners of a bank in that town.

The Twin Falls Land & Water Company was incorporated, on September 3, to build and operate the irrigation system. A. K. Steunenberg and Lowell, on the one hand, and Perrine and Stanley Milner on the other, controlled an equal number of shares. A fifth incorporator, Frank Knox — president of a Salt Lake City bank that Milner was associated with — was given a token number of shares, to enable him to break any tie votes. Milner was named president of the corporation.

To help finance his shares, Perrine conveyed to the company his water rights at The Cedars, as well as some property he owned on the river in Hagerman Valley.

After incorporation, an initial survey was done of the lands to be irrigated, by Lowell, Perrine, Ross and G. W. Burton, Perrine's uncle. On October 12, Lowell presented the state land board with a formal proposal for construction of the irrigation works, and an application for a Carey Act withdrawal of some 248,000 acres. The proposal called for a single canal on each side of the river. The land board accepted the documents and one of its members — the governor himself — went to the U. S. Land Office in Hailey, a few days later, to submit the application.

Not long before the application was made, the U. S. General Land Office had temporarily set aside, for a proposed "Shoshone Falls National Park," the four townships which straddle the Snake River and encompass the Blue Lakes, Shoshone Falls and the Twin Falls. Prominent people had been

The Assassination of Frank Steunenberg

Six years after leaving office, on December 30, 1905, Governor Steunenberg was killed by an assassin's bomb attached to the gate of his home in Caldwell.

His killing was thought to have been an act of revenge for his tough actions, in 1899, to restore order, after striking miners had committed acts of violence in the Coeur d'Alene mining district.

Harry Orchard confessed to committing the crime, and implicated several officials of the Western Federation of Miners in the plot.

In a sensational trial, which pitted the famed Clarence Darrow against prosecutor William Borah, the union leaders were acquitted. Harry Orchard was found guilty and spent the rest of his life in prison.

recommending establishment of such a park for a decade, and a petition to do so had been filed with the Secretary of Interior two years earlier.

The Land and Water Company and the state had known about the park proposal when they made the application, so a proviso had been added which allowed the company to withdraw if the order was to be made permanent.

The state engineer, D. W. Ross, quickly enlisted the help of Idaho's congressional delegation and, in November, Governor Steunenberg wrote a fervent letter asking the Secretary of Interior to reverse the withdrawal. He argued that such a park would prevent using the vast water power available in the area, and would take 50,000 acres right out of the heart of the tract to be irrigated, which would provide homes for thousands of new Idahoans.

A few months later, after his term as governor had expired, Steunenberg went personally to the nation's capital and managed to get the park proposal killed, and the Twin Falls segregation approved. Formal articles of agreement were made on July 1, 1901, between the Secretary of Interior and the state.

After clearing this roadblock, the Land & Water Company hired, as its chief engineer, Andrew J. Wiley, who had been involved in many major hydroelectric and irrigation projects in Idaho and throughout the world. Under his supervision, a more detailed survey of the dam site and canal lines was made.

The segregation was found to encompass a total of 244,925 acres, of which 220,442 were irrigable. Of the irrigable land, 26,244 acres were on the north side. The cost of the dam, and the canals on both sides of the river was estimated to be $1,519,945, which far exceeded the money available to the stockholders of the Land and Water Company.

And so, with a price tag in hand, the effort began to find the necessary capital for what, at the time, would be the largest irrigation project in the United States.

Stanley Milner

Finding The Money

—Buhl and Kimberly Take Control of Irrigation Project—

In July, 1901, after getting the federal government to kill the Shoshone National Park proposal, and approve their great irrigation project, the incorporators of the Twin Falls Land and Water Company thought that only one obstacle remained in the way of starting construction — they had to find the money.

This would seem to have been no problem for Ira Perrine — a man who, the previous year, had sent some of his Blue Lakes-grown French Petite Plums to Paris to be entered in the great *Exposition Universelle,* and had won a gold medal for his entry.

But Perrine, and a certain Harry L. Hollister would come close to scuttling the irrigation project.

Hollister was a Chicagoan who brokered and managed mining properties. He and Perrine wanted to build a power plant, at Shoshone Falls, that could provide electricity to the mines and towns of the Wood River Valley, and to a new town to be established at the falls. Their grandiose scheme called for using water brought down through a 400-foot tunnel from above the falls, where Perrine had recently claimed 1,000 cubic feet per second of water.

Harry L. Hollister in 1907.

But he and Hollister needed a 2.6-acre parcel on the Snake's north bank for their tunnel and power plant. The property was part of a state-owned section and, when the pair asked the Idaho land board to sell it, the attorney-general accused them of trying to obtain the valuable acreage "for a mere song," so they might use it for "stock jobbing in the east."

Because of the attorney-general's opposition, Hollister and Perrine weren't able to convince the land board to sell the parcel, but they did obtain a five-year lease on it, with an option to buy after the project was well underway.

With the lease in hand, token work was begun on the tunnel in early January 1902. Shortly thereafter, a group of New York capitalists, led by a certain William Dowe, agreed to organize a company to operate Hollister's mines and ore stamps, and to finance the power plant.

But, at the same time, a legal battle began between the Hollister interests and those who owned the hotel at the falls, and much of the surrounding property. Because of irregular survey lines, the hotel owners claimed that the parcel for the proposed power plant belonged to them, not the state.

Fearful of the effect the dispute might have on his lease, Hollister tried another method to gain control of the property. In early February, he sued the state and the hotel owners, seeking to condemn the 2.6 acres as a public right-of way, by proceedings in eminent domain. Both sides hunkered down for a lengthy legal battle.

The power plant scheme had greatly disturbed state engineer D. W. Ross, and the men he had gotten involved in the Land and Water Company — A. K. Steunenberg and James Lowell — who thought such a enterprise might encroach on the company's irrigation water right. The company's

Frank Buhl and Ira B. Perrine, ca. 1909.

chief engineer, Andrew Wiley, agreed with them, pointing out that, in late summer, there was normally only 2,090 c.f.s. of water in the Snake, and that, for two years in a row, it had gone completely dry above American Falls. Since one enterprise or the other was sure to suffer in low water years, Wiley suggested building the power plant farther downstream, where the river was recharged by large springs.

But, Perrine and Stanley Milner, the president of the Land and Water Company, had another solution in mind. Sometime in April, they got Dowe and Hollister's tentative agreement to finance both projects as a united venture.

Perhaps to force Steunenberg and Lowell out of the company, Milner called a directors' meeting in June 1902 at which an assessment of 10 cents per share was levied upon corporation stock. But Steunenberg and Lowell — concerned about the potential water rights conflict — weren't willing to pay. They proposed instead that the corporation be dissolved and the irrigation works be built by the federal government. This idea was defeated by the votes of Perrine, Milner and the banker, Frank Knox.

And so, on July 7, the "holdbacks" — Steunenberg and Lowell — sold their stock to Stanley Milner for $101,000. The new arrangement gave Milner 89.9 percent of the shares.

By this time, the state land board was becoming impatient with the delay in getting the irrigation project started, so Stanley Milner was asked to appear before them on November 8, and show that the company was ready to begin work, or show cause why the project shouldn't be canceled.

The request greatly alarmed Milner because by then — due to the ongoing legal fight over the power plant property — William Dowe and his monied associates had reneged on their promise to finance the power and irrigation projects.

So Milner stalled. After asking the board for a short delay, he had some preliminary work started at The Cedars, while he and Perrine looked about desperately for someone to back the irrigation project.

As it happened, they got lucky. Perrine had previously offered a broker of mining properties, Witcher Jones, a handsome commission if he could find someone to put up the required capital. In mid-November, Frank H. Buhl and his wife, Julia, came to Salt Lake City to look at one of Jones' offerings. Buhl and his friend, Peter L. Kimberly, were both awash in cash, having recently sold their steel mills in Sharon, Pennsylvania.

By the time Buhl arrived in Utah, the mining property had been sold, so, when he asked if there were any other good investments, Jones told him about the Twin Falls irrigation project. Buhl agreed to look it over, and, within a few hours, he and his wife were hustled off to Idaho, along with Stanley Milner and his wife, Truth, and Walter G. Filer.

Filer was an engineer who had managed various Utah mining enterprises for Buhl, Kimberly and Milner. As a former Pennsylvania resident, he had known Buhl since boyhood.

Upon their arrival at Shoshone, on November 24, Burt Perrine showed them his prosperous Blue

Modest Paul Bickel

At the time Paul S. A. Bickel was hired, he was one of the most prominent civil engineers in the West. An Ohio native, he had studied in Germany and subsequently been involved in wide variety of projects. Most recently he had worked as Montana's state engineer.

Though Bickel was an irrigation whiz, he had the reputation of being a modest, approachable man. Unlike many engineers, said one friend, he never wore "spats, leggings, corduroy coats or sombrero hats," nor did he "strut about with a roll of blue prints under his arm and talk of what 'we' are doing."

Lakes farm and the proposed tract, driving across 130 miles of rough sagebrush plain over the next four days.

Convinced, Buhl signed a preliminary agreement to pony up the estimated $1.5 million to build the irrigation works. In return, he would get 62.5 percent of the Land and Water Company stock. Buhl would become president of the company, and Milner vice-president. Buhl would give Walter Filer an interest in the company, and name him general manager. Since Buhl also wanted his own man as chief engineer, Filer recommended his friend Paul Bickel, who was hired.

Buhl left for Pennsylvania, tasking Filer to confirm previous surveys and investigate titles, after which a final contract would be made. On his way home, Buhl met with Peter Kimberly, who agreed to join him in the venture.

Chief Engineer Bickel took a team of surveyors across the snow-covered desert to The Cedars and, from December 7 through 25, confirmed the canal lines. His findings were approved, and a final contract with the state was signed on January 1, 1903. The document made it clear that the irrigation water right was superior to that of the power company. The signing was celebrated by feasting on some persimmons Perrine had grown at the Blue Lakes.

On January 14, the agreement with Buhl was formalized and construction of the great irrigation system was begun.

The Surveyor

– The Busy Lives of John and Anna Hayes –

It was the summer of 1903 when John Hayes got his first glimpse of Anna Hansen, the girl he would marry. She was on vacation from her studies at Albion State Normal School, and he was a member of a survey crew camped near the Rock Creek store run by her father, John Hansen.

Anna caught Hayes' attention by baking a treat for one of the other surveyors — a pie that looked delicious, but which, in the eating, proved to be filled with oats and chaff. Hayes thought the seventeen-year old prankster was "a girl with real spirit!"

Unlike his friend, Paul Bickel, Hayes had only a high school education. He learned his civil engineering skills on the job, beginning in 1893 at Cripple Creek Colorado, as a chain man, at age sixteen.

He had been working for several years on a survey of Montana public lands, when, in December, 1902, Bickel brought him to Milner to help examine the proposed Twin Falls project for financier Frank Buhl.

After Buhl and Peter Kimberly took control of the Twin Falls Land and Water Company, in 1903, six surveying parties went into the field to map the tract topography and lay out lines for the Milner dam, the canals and laterals. For $80 per month, Hayes operated the transit in T. A. Perkins' fourteen-man group.

During the next few years, Perkins' crew worked the tract from

Surveyors' camp at Rock Creek Crossing, August 1903. John Hayes standing at left. *(Courtesy Robert K. Maddock Jr.)*

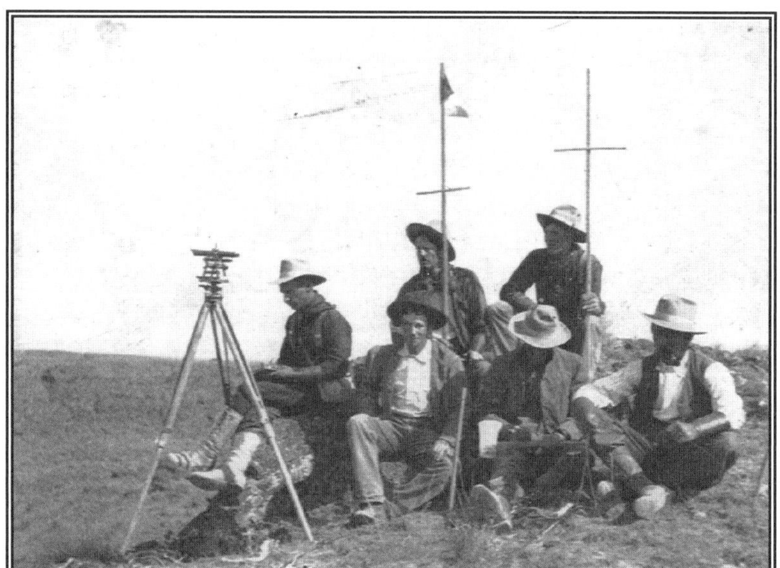

Transit party of engineering Crew No. 3, August 1903. John Hayes at far left. *(Courtesy Robert K. Maddock Jr.)*

Anna Hansen, 1904 graduate of Albion Normal. *(Courtesy Robt. K. Maddock Jr.)*

one end to the other. They laid out the Dry Creek (Murtaugh Lake) dam, located the four corners of the section within which the town of Twin Falls would be built; and they helped find routes for the Low Line and High Line canals.

The survey crews slept and ate in tents, competing for space with the many rattlesnakes. Most firewood and water had to be purchased and hauled to their ever-changing campsites. Trips were made to the "bridge" (Milner), to Stricker's, or to John Hansen's store for mail and supplies. Aside from drinking and gambling, the surveyors' only diversion was an occasional dance at Rock Creek.

Beginning in April, 1904, Hayes led the crew that surveyed the Twin Falls town site. As each phase of the work was finished, he eagerly jumped in his new buggy and carried the latest revision of the plat to the Cassia County seat at Albion. There, after recording the document, he would take up his courtship of Anna, who had just graduated from the Normal School, and soon would become an assistant professor of English and mathematics.

In March of 1905, with the first water about to be turned into the canal, and his work on the south side tract winding down, Hayes and another man opened the Liberal Cigar Store in Twin Falls. Hayes sold his interest in the store within just a few months, but, in another, longer-lasting business venture, he and Richard Cookinham became partners in the Twin Falls Engineering Company. Hayes was busy during this period surveying the town sites of Burley, Hansen, Filer and Buhl, and supervising the engineering work on the original Rock Creek bridge at the foot of Shoshone Street.

John and Anna were married on Christmas Day, 1905 in the Hansen home at Rock Creek. Concerning the marriage, one of Hayes' bachelor friends remarked that the surveyor would henceforth keep "square with the world," even though he had earlier laid out the town of Twin Falls off-kilter. After the wedding, Anna resigned her teaching job and the couple set up housekeeping in Twin Falls.

Over the next twenty years they brought four children into the world. The first, Winifred, was born December 1, 1906 while her father was working on the Fish Creek reservoir project near Carey.

We have Hayes to thank — or curse — for conceiving the idea for the Salmon Tract, which

Anna Hansen Hayes

Anna Hayes, like her husband, was a person of many accomplishments. She was very active in the Parent-Teachers Association, serving as national president of that organization in the late 1940s. In 1950, during the occupation of Japan, she served as a special consultant to the Japanese P. T. A.

Anna also served four years as state director of the Foundation for Infantile Paralysis.

Throughout her life, Mrs. Hayes wrote numerous articles and poems for newspapers and magazines, and was the author of three books – a poetry collection entitled *Lure of the Trail,* and two stories about life at Rock Creek, *The Adventures of Hedvig and Lollie* and *Buckskin and Smoke.*

lies south of the High Line Canal, and is occasionally tight on available water. While surveying the High Line in 1904, he found an ideal place for a dam on Salmon Falls Creek. Although others, including Ira Perrine, reaped most of the profits from this project, Hayes did the lion's share of the engineering work, in 1907, on the dam, tunnel and canals. Following this, he helped survey the north side tract and the towns of Jerome, Wendell and Hollister.

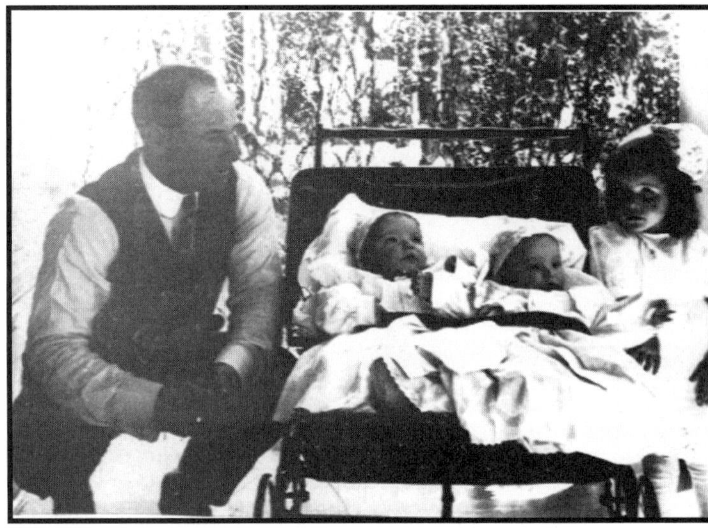

John Hayes and his children, ca. 1911.
(Courtesy Robert. K. Maddox Jr.)

The Hayes family moved to Denver in 1909 but returned to Twin Falls twenty years later, when John was offered a job managing the south side canal company. Subsequently, he served as city engineer and county surveyor. While in private practice he designed and mapped many of the city's subdivisions. After a very busy life, he died in 1962, at age 84.

Today, everywhere one looks in Magic Valley, can be seen the hand of Johnny Hayes, but his name adorns no school, street or public building. Of all the men involved in creating our towns and great irrigation projects, he, perhaps, has received the least credit due.

Ready for an outing, 1904. Back, from left: John Hayes, Henry Hansen, Sidney Strong. Seated: Anna Hansen, Ella Robinson. *(Courtesy Robert K. Maddock, Jr.)*

The First Opening

– There Was No "Oklahoma Land Rush" For Twin Falls Land –

Milner Bridge with town in background

The initial "opening" of the Twin Falls tract, offering 60,000 acres of land and water shares for sale, was a miserable flop. Some old timers who have written about it maintain that, after learning that only 3,800 acres had been taken up, Frank Buhl, president of the Twin Falls Land and Water Company, left for the East, muttering instructions to his general manager, Walter Filer, to meet him in Chicago, and to bring along "that farmer, Perrine," who had claimed there would be a great demand for the land.

The opening had taken place on July 1, 1903 at the Opera House in Shoshone after elaborate preparations were made for an expected horde of land-seekers. Special railroad fares were offered; a fleet of buggies stood ready to take buyers out to examine the tract, and company agents were stationed at strategic points, ready to furnish information.

Promptly at 10 A. M. that day, little Jessie McF all, the daughter of a Shoshone businessman, was led forward to select the name of the person to get first crack at the land. She inserted her little hand into the box and withdrew the application of Joseph M. Hollister, the deputy assessor of Ada county.

A year hence, there would be an abundance of people filing on the Twin Falls land – schoolteachers, Eastern capitalists, employees of the Land and Water Company and its contractors; railroad men, and farmers from all over the country. But, at that first drawing, only 57 individuals put their names in the box and, of these, only 28 actually filed with the state land board agent on the scene. And, nine of these entrymen were Burt Perrine's family or associates.

Joe Hollister filed on a quarter section fronting on what would become Falls Avenue, three miles east of Blue Lakes Boulevard. Most of the entrants chose land near the site which Land and Water Company officials had initially selected for a town to be known as "Twin Falls." This supposed townsite would be situated between the Shoshone and Twin falls, about a mile south of the river. It was believed that, with the coming of the railroad, this town would become a popular resort, even should the commercial center be located elsewhere.

In part, the poor turnout was due to Frank Buhl's failure to adequately publicize the project. He is said to have boasted that not a dollar had been spent for advertising, because he thought the news of this grand opportunity to take up land would spread by word of mouth.

There was also competition, at the time, from the government's Minidoka irrigation project, where

Flume and power house downstream of Milner Dam.

What The Land Cost

Each entryman could buy, from the state, up to 160 acres at 50 cents per acre. Half the cost had to be paid up front, and the rest when "final proof" was made.

For each acre, the entryman also had to buy one water share in the canal system, from the Land & Water Company, at $25 per share. Three dollars of this had to be paid at the time of filing, with the balance in annual installments. Thus, a settler could get started for a total cost of $3.25 an acre.

First Family

The first family to take up residence on the tract was that of Lyman E. "Ene" Prothero, and his wife Annie.

Former residents of Baraboo, Wisconsin, the Protheros pitched a tent on August 2, 1903 and began clearing their 80-acre parcel, located about two miles south of Kimberly.

By Thanksgiving, a house of sorts had been erected and, within its walls, the Protheros served dinner for the Robert Brose family of Rock Creek.

lands could be had for a cheaper price. But, later, most Minidoka buyers would come to regret their choice, when their fees went up, and the delivery of water was delayed for many months.

But, the biggest reason for the meager sale of Twin Falls land was the failure to get the project off to a good start and give people confidence that it was going to happen. Walter Filer had said that Milner dam and at least 25 miles of canal would be finished in time for the 1904 growing season but, by the time of the drawing, nearly nothing had been done.

Frank Buhl had agreed to finance the project, and had taken control of the Land and Water Company in mid-January 1903, but a contract for building the dam and headworks hadn't been signed until February 21. Worse, the contract to build the south side canal wasn't let until March 10.

Early on, the contractors — Faris and Kesl of Nampa — had built a bridge across the Snake, but, before any heavy work could begin on the dam and its approaches, they also had to build a hydroelectric plant, to supply power for running a narrow-gauge railway, an overhead cableway and other equipment. Downstream of Milner dam, water would be diverted into a half-mile long flume to drive a 500-horsepower electric generator. By the time of the first land drawing, the dam and flume were complete, but the generator had not yet arrived.

After the drawing, Frank Buhl returned East, a thoroughly disillusioned man. Sometime later, "that farmer," Perrine, met with Buhl, as ordered. At the meeting Perrine and C. B. Hurtt — a successful Boise publicist and land promoter — convinced Buhl to give them a contract to sell all of the water rights under the canal system. They agreed to sell 25,000 acres worth of water shares per year, for a commission of ten percent of the sales price. With this contract in their pockets, Perrine and Hurtt set about organizing the Twin Falls Investment Company, which, later — with the use of a vigorous advertising campaign and the efforts of a fellow known as "Cyclone Bob" McCollum — would be highly successful in finding buyers.

Not By a Dam Site

—Milner, The Town That Vanished—

Young baseball enthusiasts of Milner, Idaho had a favorite cheer during the early, booming, days of their little town's existence. They would shout:

"Dam! Dam! Milner by the dam!
Are we in it? I should smile!
And we'll be in it a dam long while!"

Everyone was certain Milner would soon become a great metropolis. "Conservative men" — so said the Boise newspaper — were predicting that it would be "a bustling city of 15,000 people" within five years.

Named in honor of Stanley Milner — one of the principal men involved in the Twin Falls irrigation project — the town was laid out, beginning in January, 1903, on the south bank of the Snake River near the site of the proposed new dam. Within just a few months it was a busy, white-tented "company town," comprising about 200 more or less permanent residents, and another 800 or so

Main Street, Milner, Idaho, ca. 1905.

Kimama

The railroad siding nearest to Milner was at Kimama, some 25 miles to the north, across a sage brush plain.

A few Oregon Short Line buildings and employees had been there since establishment of the line in 1883, but the start of the great irrigation project made Kimama blossom, if only briefly. Until the new Twin Falls branch railroad was built as far as Milner, it was the transshipment point for everything needed to build the dam and canals.

As at Milner, the Perrine clan was quick to see the opportunities at the little watering stop, putting in a Perrine and Burton general store, as well as a restaurant and a lodging house.

Ira Perrine obtained a contract for moving supplies to the dam, and he installed brother Silas at Kimama to oversee the operation. Silas also ran a livery stable, and was the postmaster and a Justice of the Peace.

workmen, who were starting to build the Twin Falls irrigation project.

The Land Water Company's two-story office and lodging building, which was finished in June, became a principal center of activity. Situated with a commanding view of the dam and river, it was both home and office to a number of officials and engineers.

James Bower — the cattleman who had originally owned most of the town site — was in business there for a time, running a hotel and a livery. There were a handful of other businesses in town, the two largest of which were the Stock Grower's Mercantile — owned by Edward Hartdegen, the recently-appointed postmaster — and the Perrine and Burton general store. The latter was owned by the brothers, Silas and William B. Perrine, and their cousins, Silas and William Burton. Will Perrine and Will Burton had just arrived a few weeks earlier from Anaconda, Montana for what was to have been a week's visit.

Milner's first social function was the dance hosted by Ed Hartdegen just before he stocked his new store with goods. The 25 women and 150 men in attendance danced to the music of two violins and a cornet and, afterward, enjoyed an elegant supper, served by Charles Sturgis in his new restaurant.

Later on, the dances were regularly held, on Friday nights, in the Milner "dance hall"—a large tent. A deputy sheriff, assigned for the Milner precinct, kept things pretty well in check, although sometimes it was necessary to relieve men of their guns before letting them come into the dance.

With completion of the south side project in 1905, the little hamlet's population dropped for a time, but another brief period of prosperity came for several years, beginning in 1907, when work on the north side canal system began. The Kuhn brothers — capitalists from Pittsburgh — organized the Twin Falls North Side Land & Water Company and bought out the Buhl-Kimberly Corporation's interests in the Milner town site and in the north side project.

Even though it was four years old, Milner was "opened" for settlement in April 1907, coincident with the first drawing for north side land. It was claimed that $45,000 worth of town lots were sold, but records show that only 33 Milner lots changed hands during the entire period from 1903 through 1909.

In early 1907, the Land & Water Company began adding rooms to the company lodging house and, by November, they were calling the place the "Riverside Inn." Though it continued to provide offices for all of the Kuhn interests, when remodeling was completed, in May 1908, the inn was opened to the general public.

It was, by any standard, a lovely place, with its 36 bedrooms, six suites with private baths, several parlors and a large dining room. Steam heat and electric lighting further added to the occupants' comfort, and the hotel's wide veranda offered them a nice view of the "Irrigation Falls" flowing over the dam.

The officials running the Kuhn enterprises included a group of Ivy League easterners who must

Riverside Inn

have thought they had been banished to the wilderness. So scarce was entertainment in Milner that occasionally, of a Friday evening, townsmen would take visitors to watch a ferocious mongrel dog, that lived at the company warehouse, fight a tough old badger named "Jack." But, eventually, even Jack lost interest in this pastime.

In order to make life bearable, the Milner Country Club was started in July 1908. The company laid out an athletic field and built a dancing pavilion at the rear of the Inn. A half-mile racetrack was laid out "for the ponies to trot on." And, using a boat named *Stella Perrine,* and others, moonlight "launch parties" were held on the reservoir, the boats traveling as far upriver as the Minidoka dam.

About this time, the town's promoters once again predicted a prosperous future. Centrally located as it was, Milner was "destined to be the playground of the Twin Falls and Minidoka projects." With the fine hotel, the country club, the "mammoth dancing pavilion, splendid boating, fishing and bathing," and the racetrack, one had only to "watch Milner grow."

But regardless of its high-toned luxuries, Milner never in its existence had some of the normal amenities of a town. Although a school was started in September 1903, with Miss Mattie Mathews teaching about 20-30 pupils, a proper schoolhouse was never provided. No churches were apparently ever built, although services were conducted in a small chapel, and Mrs. S. F. Kesl held Sunday school in her home. The town never had a real jail, a sewer system or more than a primitive lighting system.

After the north side project was completed, Milner began to die. In 1914, the Milner State Bank and other businesses began leaving, one by one. The Inn was demolished in 1923, and today a home used by the caretaker for Milner Dam stands on its former site. All that remains of the short-lived village is the concrete shell of the bank's vault.

Danger Ever Present in Building Milner Dam

Overhead cableway dumping rocks; town of Milner in the background.

On a cold winter day in November, 1903, Matt Myers was thawing out some of the granular dynamite known as "Giant Powder," at the site of Milner dam, when the explosive blew up in his face, sending him flying, and tearing a great hole in his head.

It was supposed that the young laborer had been smoking at the time, and had dropped some sparks. The dam contractors, Faris and Kesl, obtained a casket and shroud from Boise for the corpse.

Myers' demise vividly illustrates just how risky building a dam was in those days. In addition to mishandling dynamite, there were a number of other ways to lose life or limb — exploding steam boilers, falling rocks, a treacherous river, and runaway horses, to name just a few.

At The Cedars, where Milner dam was built, there were formerly three channels of the river, separated by two large islands. The dam was built in three sections, to join the islands to the river banks and raise the water level to flow into the south side canal. The dams were designed by Paul Bickel, with the help of J. D. Schuyler, a noted hydraulic engineer from California.

The dam sections were built with a central core of heavy planking, anchored to the river bed. Embankments of loose rock were laid on both sides of the core and, on the upstream side, the rock was filled and banked with puddled dirt, then paved with rock rip-rap.

In its natural state, except at high water, the river ran only in the north channel. Faris and Kesl's construction plan called for isolating the

A Talented Engineer

Faris and Kesl Engineering Company was chosen in February 1903 to build Milner Dam. Robert W. Faris, senior member of the firm, was a childhood friend of Mrs. I. B. Perrine.

From 1899 through 1901 Faris was involved in the design and building of a power plant at American Falls, which supplied electricity to Pocatello. Later he did most of the engineering work for the Shoshone Falls power plant.

He also built a suspension bridge at the Blue Lakes ranch, which, when completed in 1902, carried a large water pipe across the river to some placer claims.

Tunnel cut through the southernmost island.

two southernmost channels, and building those sections of the dam before the river rose. A large tunnel would be cut through the south island and, when the river subsided in the fall, its entire flow would be diverted through that tunnel while the northern dam was put in. Upon completion — scheduled for early 1904 — steel gates would permanently shut off the tunnel and water flow would thenceforth be regulated by 99 spillway gates on top of the south island.

As noted in an earlier article, not much was accomplished on the dams in the six months following the Buhl-Kimberly Corporation's takeover of the project in January, 1903. The delay was largely due to difficulties in getting a water-powered electric generating plant installed with which to run a narrow-gauge railway, an overhead cableway, and other heavy machinery. While waiting, men and dozens of teams of horses were used to excavate foundations for the first two sections of the dam, and to begin driving the tunnel through the south island.

Finally, in late August, the power plant was ready, and work began in earnest. Most of the rock and dirt used in the dams was taken from the head of the canal. Electric drills and explosives were used to break up the rock, which was loaded by a large steam shovel into skips (dump cars). Then an electric locomotive hauled the cars on a narrow gauge railway to the dam site, where electric hoists from an overhead cableway picked them up, transported them to the desired place, and dumped them. At other points along the head of the canal, men and teams loaded dirt into dump cars which was also sent to the dam for sluicing into place.

Workers riding a skip on the overhead cableway.

After arc lights were rigged in the work area, Milner resounded, both day and night, to the thunder of dynamite blasts, the clatter of electric drills and the singing of hoists and cableways. Faris and Kesl optimistically declared that all they needed now was 100 working days to complete the dam.

But a cold winter slowed the work and, by late April 1904, as the river rose, the Land and Water

Company was forced to admit that there would be no irrigation water on the land that year, as had been planned. Construction of the third section of the dam would not begin until the river dropped in the fall and, as events proved, the entire structure would not be completed and begin diverting water until March 2, 1905.

Milner Dam, shortly after its completion.

Built on a Bias

—Those Cockeyed Twin Falls Streets—

Who among us hasn't cursed the street layout in the oldest section of Twin Falls? In a region where everything is on the square, with roads running mostly north-south and east-west, our downtown's off-center layout is both perplexing to newcomers, and an annoyance to old-timers. And, because of this off-kilter design, we have a confusing street-naming system that has bewildered citizens since the very beginning.

Just who devised this maze? A fellow named John E. Hayes is to blame — maybe. Hayes was one of that corps of engineers who, under Paul Bickel, helped survey the canals and topography of the Twin Falls Land and Water Company's irrigation project, beginning in January of 1903.

Early on, several locations had been chosen for the principal city of the tract, but all were discarded because railroad officials didn't think them suitable for their intended route.

John Hayes at the intersection of Shoshone and Main. *(Courtesy of Robert Maddock Jr.)*

Finally, Bickel selected Section 16 of Township 10 South, Range 17 East of the Boise meridian. The railroad liked the choice, and the parcel had the added advantage of being near the center of the project. It was also a state school section, which allowed its purchase without meeting any federal requirements.

In August 1903, while his survey crew was camped where the old Oregon Trail crossed Rock Creek, Johnny Hayes located the four corners of Section 16. But, no action was taken, for months, to buy the parcel or to survey it in detail. Only when it became apparent— in late March 1904 — that other promoters were planning to lay out a town near Shoshone Falls, to be called "Twin Falls," did the company buy Section 16.

A few days later, Bickel summoned Hayes to Perrine's Blue Lakes Ranch, and told him to turn over his regular crew to someone else, hire some helpers and begin surveying the town.

Hayes recruited two land seekers who were staying at Blue Lakes and, beginning about April 5, took them every day across Snake River on Perrine's ferry. There, they climbed out of the canyon on the rough wagon road built several years earlier, and walked to what is now the center of Twin Falls, carrying their equipment and food.

They laid out four blocks at the center of the section, on the same 45-degree bias that exists today.

Ira Perrine and Tom Costello of Twin Falls Investment Co. look over the prospective Twin Falls town site.

The thoroughfares at the center were named Main Street and Shoshone Avenue

A plat of the four blocks was drawn and, on May 12, Hayes drove a buggy to the Cassia County seat at Albion, to register it and secure the name "Twin Falls." He may have also stopped off at the Normal School, where his bride-to-be, Anna Hansen, was in residence.

By late May, a full surveying crew was at work, and their tents, pitched near the present East Five Points, became home for Twin Falls' first residents.

On June 16, the Twin Falls Investment Company came into being, to which the Land & Water Company gave a contract to sell the south side water rights and to develop and sell the townsite. Buying equal shares were Clarence B. Hurtt of Boise, Ira Perrine and seven others.

Probably as a publicity stunt, Hurtt brought a renowned architect, Emmanuel Masqueray, to the tract in late June. French-born, Masqueray had just completed three years as chief of grounds design for the Louisiana Purchase Centennial Exposition in St. Louis, where Hurtt had been Idaho's representative. The two arrived in the company of Idaho governor, John T. Morrison and Mark Bennett, a publicity man who had been manager of the press bureau at that same great fair.

After the group visited with Ira Perrine and checked on the town site progress, Masqueray left the area on July 5, never to return. But, Masqueray's visit had added a touch of class to the town, and soon Mark Bennett was sending press releases across the country claiming that the noted architect had "laid out" Twin Falls.

But, this is unlikely. At the time of the Frenchman's visit, the first plat had already been recorded and several buildings were going up.

Another plat was filed on August 19, adding fourteen blocks to the original four. A third filing was made in October, encompassing 36 blocks, and a fourth and final plat was recorded December 16, comprising the entire section.

So, why was Twin Falls built on a bias? Shortly after the surveys, John Hayes said it was so the prevailing winds would blow across streets instead of sweeping them from end to end. The offset configuration would also allow easier irrigation of the townsite. In later years, Hayes said the layout had been decided upon by Bickel, and he gave more reasons for the decision — the sun would shine in every room sometime during the day, and drainage of the sewer system would be made easier.

But the real reason may the one given by the Boise *Statesman* at the time — so that Shoshone Avenue might lead directly to Shoshone Falls, as it would have, had it ever been extended.

The original street naming system had the **avenues** parallel to Main **Street**, and **streets** parallel to Shoshone **Avenue.** The numbers for avenues and streets began at the northeast and northwest corners of the section, respectively. Because of Shoshone Avenue, there was no 8th Street, and because of Main Street there was no 11th Avenue.

As soon as this system was created, citizens began complaining, relentlessly, about it. George

Fraser, editor of the Twin Falls *News*, called it an "abomination," so bad that a stranger to the town would have to hire an engineer in order to locate a house. Even residents had trouble remembering what street *they* lived on. Only Mark Murtaugh defended the system, blaming the confusion on the "ignorance of the people."

And so, in May 1908, the city adopted the present system. Main Street became an Avenue, just like the thoroughfares parallel to it, and Shoshone Avenue became Shoshone Street, just like those running parallel to it. And, thoroughfares were numbered outward from Main and Shoshone, with the first being called Second Street/Avenue. In addition, each was described as being North, South, East or West, according to the quadrant in which it lay from the starting point.

Fraser declared that the new system would be an improvement and, if it wasn't, the people could "have it changed." And, indeed, that second abominable system, which has existed for 93 years, may soon be replaced — one hopes for the last time. A committee representing downtown businesses has recommended the naming of all the streets but Shoshone (but not the avenues) after notable citizens of Twin Falls.

The name of Johnny Hayes — who helped start all the confusion — is conspicuously absent from the new list.**

** Subsequent to publication of the first edition of this book, the **streets** in question were named after surrounding towns and landmarks, not after notable citizens.

"Cyclone Bob"

—Selling The Land—

Ira Burton Perrine may have been the father of the Twin Falls tract, but it was "Cyclone Bob" McCollum who got it sold.

A warm-hearted, hail-fellow-well-met type of man, he was a natural salesman. It was said that when he went to Boise for a hernia operation, the first thing he did upon reviving from the anesthetic was to sell a Twin Falls lot to the surgeon, and this made Bob "as cheerful as a boy with a new pair of red top boots."

Investment Company building on August 1, 1904. Bob McCollum on the porch, with John Hayes' survey party nearby.

In 1896, Robert M. McCollum — a printer from New York — moved to Shoshone, Idaho, where he became publisher of the *Shoshone Journal*. With him he brought Alice, his wife of 14 years, and their young daughter, Elva.

Within months of their arrival in Idaho, the family began a warm and long-lasting friendship with Ira and Hortense Perrine. McCollum sold his newspaper in 1899 to George Eveleth, husband of Mrs. Perrine's sister, Stella, and, for the next few years he was involved in a variety of new activities.

For a time he drove one of Perrine's stage coaches carrying passengers from the railhead at Shoshone to the Blue Lakes and Shoshone Falls. He also worked a placer claim on the Snake River, using a steam-powered gold-saving device. In 1902 he served as secretary of the Republican State Central Committee, under Frank R. Gooding—who later became Idaho's governor.

Sometime in 1899 Perrine had become interested in irrigating a huge tract of desert land using water taken out of the Snake River at "The Cedars," the site of present-day Milner Dam. One of Perrine's biographers has said that McCollum was the first person Perrine told of the scheme. Perrine is supposed to have offered McCollum a share in the deal if he would help obtain capital for the project, but when McCollum saw the powerful river surging down the canyon, he told Perrine he was a "damned fool" to think that the river could be harnessed, and that he, McCollum, wanted nothing to do with it.

Nevertheless, from 1903 onward, McCollum was fully involved in Perrine's expansive ventures. He became general agent for Perrine's stage line, and secretary of his Shoshone Falls Power Company, which was struggling to build a hydroelectric plant at the falls.

Meet Me in St. Louis, Louis...

In the year of Twin Falls's birth, 1904, the Louisiana Purchase Exposition was held at St. Louis. People from all 45 states flocked to see the marvels on display at this grand world's fair that celebrated a golden age of discovery and invention.

Just a few months earlier the Wright Brothers' had succeeded in flying their heavier-than-air machine at Kitty Hawk. Automobiles, motion pictures, and the "wonder drug" aspirin had come into existence only in the last 10 years.

At the fair, Thomas Edison helped set up the Palace of Electricity that showcased the uses for this relatively-new and wondrous tool.

It was also at the fair that many Americans saw their first motion picture, and were able to see how wireless telegraphy (radio) and the telephone allowed communication across thousands of miles.

Not many of Idaho's 200,000 citizens got to St. Louis, but the state won 56 medals for various displays, including a gold medal for its ag exhibit.

On June 16, 1904 McCollum also became secretary and sales manager for Burt Perrine and C. B. Hurtt's Twin Falls Investment Company, which was incorporated that date to sell the lands and water shares of the irrigation project and the Twin Falls town lots. At the time, only 15,000 of the 60,000 acres thus far opened had been taken up by settlers.

Bob McCollum left his family in Shoshone and, in late June, came to the future site of Twin Falls, equipped only with a bunch of blueprints, a water bag, and his voice and vitality. One pioneer of the town would later remark — with only slight exaggeration — that, at the time, "the population of Twin Falls consisted of 1,000 jackrabbits, Robert M. McCollum and John E. Hayes."

Hayes and a crew were about half finished with a survey of the town site, and other men were clearing it of sagebrush. Mark Murtaugh, assistant general manager of the Land & Water Company, had just started burning brick using wood hauled from the south hills.

Thirsts were slacked at first by use of a wagon that dispensed water at 50 cents per barrel. Later, a 12- foot wide, shallow well was dug in Rock Creek Canyon. The only contact with the outside world was via Burt Perrine's daily stagecoaches that ran from Rock Creek and Shoshone, and a single telephone line connected to the Blue Lakes.

McCollum lived in a tent while a simple frame building — the first to go up in Twin Falls — was being built on Main Street. He occupied this building and, on July 27, took up his sales work in earnest; four days later he sold the first Twin Falls lot to Scott Hazen, who soon began erecting the town's first hotel, the Twin Falls Rooming House. Initially, lots were priced from $100 to $1,000. Buyers had to pay at least one-fourth in cash, and the balance in three annual payments with interest at 6 per cent.

At first, the powers of persuasion of "Cyclone Bob" — as he came to be known — were sorely taxed and people thought he was slightly "dippy" when he told them of the future of Twin Falls and the surrounding farmland. But his vigor and skill prevailed and, under his direction, about three million dollars worth of property and water rights was disposed of during the succeeding 23 months.

McCollum (seated) and Hayes.

Digging the "Big Ditch"

—It Was Dirty, Backbreaking Work—

Travelers from Milner to the Blue Lakes in the summer of 1904 could look across the Snake River canyon and see a long streak of white earth trending southwestwardly across the desert, punctuated, at intervals, by a dozen or more great clouds of dust, where contractors were working on the Twin Falls Land and Water Company's "big ditch."

Working on the main stem of the canal.

Near each center of activity along the canal's route were clusters of white tents that were home to the many laborers working on the project.

Nearer to Shoshone Falls, an occasional settler's shack could be seen jutting up from the sagebrush and, here and there, ground had been prepared for the irrigation water that had been expected, but not delivered, that spring.

Canal workers camp.

Today the Twin Falls south side system waters some 202,000 acres of highly-productive land, through its many miles of canals and feeder laterals. Original plans had called for a single, 65 mile long canal, but, by January, 1904, or earlier, the idea of high and low line canals had been conceived.

The Nelson Bennett Company of Tacoma, one of the largest firms in the West, was the principal contractor on the main canal system. Bennett and a number of smaller outfits used hundreds of teams of horses to dig and scrape out the canals

Rock Creek Siphon, March 1, 1905. Standing, left to right: J. D. Schuyler (consulting engineer), Walter Filer and Frank Buhl. Peter Kimberly is seated.

and laterals, but much of the hard, dirty work fell to men using picks and shovels. The only powered tools to be used were pneumatic drills and an "automobile scraper," so balky it had to be operated by a factory representative.

Digging the south side canals and laterals began in earnest in April 1903. Although the system would not be completed to everyone's satisfaction for another five years, water was available on about 32,000 acres for the 1905 growing season, and on the majority of the tract starting in 1907.

The contractors were required to give prospective settlers hiring preference, so quite a few of them signed on as craftsmen or laborers on the grading teams, while waiting for the water to come onto their land. There were also many Greek and Italian immigrants working on the project.

Wages, by today's standards, were very low. A common laborer might earn $2.50 for a ten-hour day; mechanics and carpenters got about $4.00, and a man with a team of horses could expect $4.50. Fortunately, the cost of living was equally low—bacon and coffee cost only about 16 cents per pound; a good meal in a restaurant could be had for 35 cents, and about a dollar would buy a pair of children's shoes.

At the peak of the work, the construction camps exceeded thirty in number. The rude tent colonies in which the workers lived were desolate places, with little shade in the summer or shelter in winter. Food for both man and beast, and other necessities,

The Siphon

The canals of the Twin Falls tract are, for the most part, unremarkable but, about four miles south of Kimberly, where the Low Line Canal intersects Rock Creek, is an unusual tunnel, called an "inverted siphon," that carries the entire canal — about 494,000 gallons per minute — under the creek.

The huge steel pipe was built by a Pennsylvania boiler works, owned by a friend of Frank Buhl, and was shipped by rail to Kimama in half-circle sections. From there, the pieces were shipped to the site in horse-drawn wagons. In all, 452 feet of pipe, weighing 100 tons, was transported.

Most of the pipe was laid in a trench cut out of solid rock of the canyon walls, with some 200 feet of it resting in a cradle of concrete on the bottom of the creek. The sections were riveted together, and reinforced with steel bands, after which the pipe was covered with earth and stone beneath the bed of the stream.

The siphon has been refurbished several times, most recently in 2000.

The Blooming Laterals

The new canals, laterals and coulees comprised a myriad of new streams, some the size of rivers, crisscrossing the region and impeding transportation.

The Twin Falls *News* aptly described one of the problems this caused:

"The blooming little lateral
Cut up the blooming road,
The blooming freighter came along,
And lost his blooming load."

Though it wasn't obliged to, beginning in July, 1905, the Twin Falls Land and Water Company offered some help in building roads and bridges along the most frequently traveled routes. But, it was years before transportation routes in the valley were adequate.

were purchased from Milner or Rock Creek merchants and ranchers and hauled in wagons to the camps. Much of the water required by the camps was transported from the Snake, where Nelson Bennett had installed an apparatus to hoist 40 gallon barrels of the precious liquid out of river, 600 feet up the incline of the canyon wall.

Because of the ever-present dust, supervisors, teamsters, mechanics and laborers alike were perpetually filthy. When it got too bad, or when the "greybacks"(lice) became unbearable, the men went for a "clean up" in Rock Creek, or at J. E. Bower's "plunge bath" at the warm springs on Dry Creek.

The workmen were mostly young and healthy. A number of cases of Rocky Mountain Spotted Fever, typhoid and smallpox were reported, but injuries were the most common medical complaint. There were three doctors in and around Milner, one of whom, E. W. Kleinman, made weekly trips down the line of the canal to see who needed care. Another, Dr. J. W. Emerson, maintained a primitive hospital in a one-room shack at Milner.

The back-breaking work and the hard, lonesome life on the irrigation project led to a lot of hard drinking in the camps and at Milner. At first, as required by state law, the only "booze factory" was out in the sagebrush, five miles from the nearest "public work." But, by early 1905, Milner had been incorporated and several saloons could be found there.

Inevitably, the drinking led to brawls, stabbings and shootings. The most bizarre incident occurred one evening, in December 1903, at a Milner restaurant, when a drunken laborer named Will Frayor pointed a revolver at his breast, and asked a girl seated nearby to pull the trigger. When the girl refused, Frayor put the gun barrel into his mouth and turned to John Hughes, a horseshoer, repeating the request. Without hesitation Hughes accommodated him, and Frayor was a dead man.

Stagecoach crossing an un-bridged coulee.

Tents in the Canyon

—Twin Falls Comes Alive —

Blue Lakes Ferry

This is believed to be Martin Berry and his shop. One of the ads on the front wall is for asbestos shoes that "wear like iron."

One of our towns' first buildings was Martin Berry's shoe shop. Built in Hailey, this little house was placed on the running gear of a wagon and hauled the 70-odd miles to Twin Falls in the dead of winter. Berry's ten-day trip in his "private [railroad] car"—as the *Shoshone Journal* called it—was a harrowing one, particularly the steep descent down the Blue Lakes grade.

The house arrived January 31, 1905, and was plunked down on Main Street. Within a week the shoemaker ran an advertisement proclaiming, "This is the man who don't refuse, to make and mend your boots and shoes." Berry also did harness repair, and sold electric soles, to "keep your feet warm."

The saga of the shoe shop demonstrates the problem of building a new town on an isolated and desolate sagebrush plain. For the first full year of Twin Falls' existence, until the railroad reached it in July 1905, every stick of lumber and much of the food had to be hauled by horse or mule-drawn wagons from the railhead at Shoshone. The crude road from there traversed 25 miles of desert, then made a steep descent to Perrine's ranch at the Blue Lakes, crossed the Snake River on Perrine's toll ferry, and then led up another narrow grade on the south side, and into the town site.

The ferry was run by a colorful

Originally the road into the south side of the canyon ran behind the waterfall caused by Perrine Coulee. Even under an umbrella travelers got a soaking.

character and close friend of Perrine's named Jake Fritz. A former prospector, crusty old Jake was invariably dubbed honorary "mayor" of every one of the several towns he lived in. He claimed to have dug the Snake River Canyon and, later, to have trained its woodpeckers as riveters on the Shoshone Falls power plant project. Jake got a haircut once a year whether he needed one or not.

It wasn't uncommon to see over 100 teams on the Shoshone-Twin Falls road at one time and, occasionally, that many were lined up waiting to cross the ferry. Nor was it unusual for wagons and horses to plunge off the road into the canyon. A drunken teamster might try to pass another wagon where there was no turnout, or a wagon's wheel locking chains might break, causing the vehicles to push horses over the side onto the rocks below. Fortunately, injuries to humans were rare.

The demand for lumber exceeded the ability to get it there but, once it was delivered to a building site, carpenters swarmed over the project like ants. The sound of hammers seemed never to cease, as one or more new buildings were completed each day. A man frequently bought a lot in the morning and turned the key in his doorway before sundown.

Keeping Clean

Personal cleanliness was a challenge for Twin Falls pioneers. For the first few years the town had only a primitive water system, laid down in the business district and to a few residences. Most people lived in small shacks or tents, without running water, bathtubs or much privacy.

There was a "great army of the unwashed"—as the newspaper put it—whose only recourse, for a time, was to rent the use of a tub at the Palace Saloon, or A. C. Huff's barbershop.

The wife of a newcomer to town is supposed to have gone to an attorney, seeking a divorce on grounds of desertion. She told the lawyer, "My husband bundled up his clothes and left home weeks ago, and I haven't heard from him since."

"Be patient madame," the attorney counseled, "he is probably waiting his turn for a bath at the barber shop."

The town's first newspaper, the Twin Falls *News*, began publishing on October 28, 1904. From a reading of its first few weekly editions, we get our earliest good look at the budding new town.

Most streets had, by this time, been cleared of sagebrush, and the outline of the town site was clearly visible. In the central district, mostly along Main Street, there were two each general stores, livery stables, restaurants, and saloons. There was also a lumber yard, a brick yard, a hardware store, meat market, blacksmith shop, rooming house and the Investment Company's real estate office.

E. B. Williams had moved a building from Milner to a lot on Main Street where, in December, he opened the City (Pioneer) Drug Store. A physician, Henry Clouchek, had begun a practice, and a Shoshone dentist, J. M Rogers, was making periodic visits to pull teeth.

Survey crews were working on grade and curb lines, but the crude streets were still deep with mud or dust, depending upon the weather. Board sidewalks were being built along Main Street and Shoshone Avenue; a small wooden school building was going up, and several denominations were beginning to hold religious services.

As the end of the year approached, Twin Falls was home to about 600 souls, including the "first baby in camp," Blanche Harrah, born on August 15 to James and Sophia Harrah. Despite it being winter, about half the population lived in a "tented city" in Rock Creek Canyon, which included a laundry and several other businesses. Some of those in the tents were waiting to go onto their farms, while many others were employed in town.

When a fierce winter gale struck on December 29, no doubt a lot of them wished they had never left their former homes. This little southern Idaho "zephyr" was powerful enough to put Perrine's ferry out of order, and require that stagecoach passengers be taken across the wind-whipped river, one at a time, in a row boat.

Teddy

Although the town was only four months old, 109 Twin Falls voters cast ballots in the November, 1904 election. Nearly all voted for Theodore Roosevelt to be President. It was as if, someone said, they "wished to express their appreciation of what the President has tried to do for irrigation."

The vigorous and popular young Roosevelt, had been in office since September 14, 1901 when William McKinley died of a wound from an assassin's bullet.

Under Roosevelt's leadership, trusts were regulated, the number of National Parks was doubled and the United States "spoke softly," but carried "a big stick."

"A King of the Desert Am I"

—The McCollums and the "Homeless Twenty"—

The McCollum residence. When built, it fronted on 7th Ave North, but was later remodeled extensively. It exists today as Ann's Eyewear Boutique at 691 Shoshone St.

It was a group that probably couldn't be gotten together in modern society. Simple and without much form, the "Homeless Twenty" sprang up as a result of the loneliness of the young engineers and businessmen who had gathered in Twin Falls, without family, to build the great irrigation project, or seek their fortunes. At the heart of the organization were Robert M. McCollum and his family, who collectively brightened these men's lives, by their generous hospitality.

McCollum, who was secretary and sales manager for the Twin Falls Investment Company, left his family in Shoshone and came to the townsite in July 1904, to begin selling town lots and farm land. Soon his wife Alice, and daughter Elva joined him to live on the dusty sagebrush desert that was becoming Twin Falls and, in about September, they began building Twin Falls' first permanent dwelling.

One day, shortly after the house was completed, Alice suggested to her husband that they entertain some of the many lonesome men in the area who were far from their homes, wives and friends, and who were living in tents or boarding houses with no place to spend their evenings. Bob agreed, remarking that the piano had dropped a tone or two in coming across the desert from Shoshone, but the boys wouldn't notice it, since the kitchen range was working just fine.

Accordingly, the next evening, February 4, 1905, eleven homesick fellows scraped the mud from their shoes on the McCollum's front steps—Martin DeLong (secretary of the Land & Water Company), Paul Bickel (chief engineer for the company), John Hayes (surveyor), C. S. Loveland (a bookkeeper), Julius Keuhn (farmer), Henry Wall (lawyer), Charles Cole (merchant), Thomas Macaulay (cigar store owner), Adelbert Sprague (pharmacist), Samuel Hamilton (lawyer), and a Mr. Keltie, occupation unknown. The group spent the evening visiting, singing, and filling up on home cooking.

In reporting the party, the Twin Falls *News* said "The cordial and hearty 'come again' at the hour of departure was followed by a...resolution on the part of each guest not to miss any similar occasion, and, thereupon, each hied himself home, positive that 'Old Time is a liar; we're twenty tonight.'"

The newspaper's quote is taken from "The Boys," a poem written by Oliver Wendell Holmes, for a reunion of the Harvard Class of 1829. The first stanza asks:

Has there any old fellow got mixed with the boys?
If there has, take him out, without making a noise.
Hang the Almanac's cheat and the Catalogue's spite!
Old Time is a liar! We're twenty tonight!

The clever poem goes on making its point that even the old, important men of the Harvard class—doctors, judges, congressmen, and such—had remained boys at heart.

Somehow, Holmes' "twenty"—an age—became the supposed number of men to have initially received the McCollums' hospitality, not just the eleven that had actually been present. A week later another, somewhat larger, gathering was held at the McCollum home, and the report of it was the first to use the term "Homeless 20."

The group remained active for several years, during which many festive parties took place, most of them in the McCollums' parlor. Sometimes, the lonesome men would swoop down, uninvited, in

The Homeless Twenty, January 17, 1906 at the McCollum residence. Top Row, left to right: P. W. Sweeney, Frank Burrington, S.T. Hamilton, C. H. Mull, Paul. S. A. Bickel, Charles M. Hill, Herbert Osburn, Charles P. Diehl. Second row: A. N. Sprague, Mrs. McCollum, R. M. McCollum Elva McCollum, Fred R. Reed, Dr. H. W. Clouchek, Dr. G. F. Baker, Harry F. Allen. Third Row: A. N. Frankell, Morgan G. Heap, Miss Gertrude Rauch, C. E. Cole; fourth row: W. W. Dunn, Henry J. Wall, T. J. Woods, George B. Fraser, John E. Hayes, J. V. Baker, T. C. McCauley. From a "flashlight photo" taken by Ernest Tacha. *(Courtesy Twin Falls Historical Society Museum)*

"true pioneer style," very late in the evening, and catch McCollum collarless and in his slippers.

As the circle widened it was sometimes necessary to gather at Frank Burrington's hotel, or at the drug store of Del Sprague and Jesse D. Butler, which was a favorite meeting place of the "old guard." At the gatherings there was, inevitably, a lot of story-swapping and boyish hi-jinks. There was usually a bit of speechifying and, occasionally, the card game "High Five" was played.

Best of all, there was always plenty of spirited music. Elva, a 1903 graduate of Shoshone High School, was a talented pianist, as was her mother. At their house, the two of them would preside at the piano—Twin Falls' first—accompanying "the boys" while they sang everything from the comic to the classical. Often, the evening would end with a favorite hymn, or the doxology. Once, Bob McCollum was given the alternative of singing "A King of the Desert Am I" or being turned out of his own house. He sang, and the roof was "bulged three inches" by the applause that followed.

There were also instrumentals, including one by a certain "Kappel Meister" Swydenski, who played a few selections on his violin, which were said to have been "so soulful that they would have brought tears from a coyote."

At one memorable conclave, Alice McCollum, "the godmother of the homeless," was toasted for her hospitality, and Burt Perrine, who was present, was praised for his foresight in founding the area. At midnight, everyone trudged through the mud to the Headquarters Restaurant for a buffet. There, the foundation of the Twin Falls Commercial Club (Chamber of Commerce) was laid, and the organization was perfected just a few months later, on July 11, 1905.

As the years passed, Bob McCollum assumed ever more responsibilities. He served as a city trustee for a year. In 1907 he became secretary for the Twin Falls North Side Investment Company and sold farms and town lots on the north side with great success. In 1908, he took charge of sales on

"Homeless 20" members march in a 1910 parade with sagebrush in hand.

the Salmon Tract.

To the great dismay of the community, all three of the McCollums died within the brief period of 33 months. Alice passed away, at age 55, on Easter Sunday, 1915 after a brief illness. Her death plunged Bob into a condition of deep melancholia from which he never recovered, and he also died, aged 57, in that first house built in Twin Falls, on December 21, 1917.

Just six weeks later, daughter Elva also expired of an unstated cause. By then, she had become the wife of Joel W. Craven, and was the mother of two children.

The McCollums lie, side by side, in the old Twin Falls cemetery. Over Alice's grave is a bronze tablet erected by 22 men of "The Homeless Twenty" with this inscription: "In loving memory of Alice Rozielle McCollum, Foster Mother of the Homeless Twenty, Pioneer Woman of Twin Falls, May Her Kindness Ever Be Remembered."

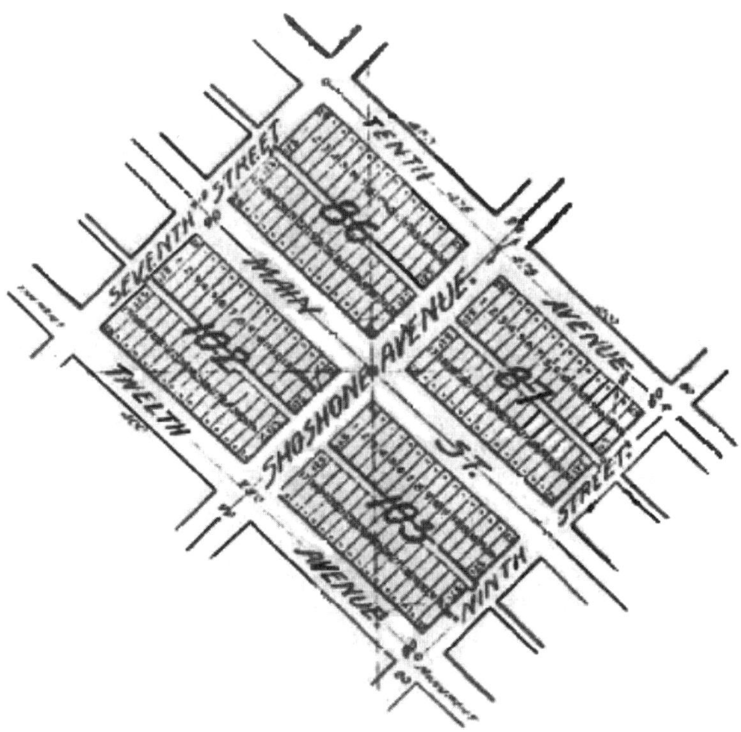

First recorded plat of Twin Falls townsite
(Author's collection)

Fighting Demon Rum

Life was tough for saloon owners in early Twin Falls. They soon learned—as the editor of the *News* put it—that "the law is mightier than the corkscrew." It was, after all, a time of fervent prohibitionist sentiment—the Prohibition Party had selected a fellow with the unfortunate name of Silas C. Swallow to run for President in 1904; the temperance crusader, Carry Nation, had been smashing saloons with her hatchet for the past four years, and the Women's Christian Temperance Union (W. C. T. U.) was, just then, the largest women's organization in the country.

And with good reason. Drunkenness and alcoholism were all too common in that era of rude living conditions and low wages.

Almost from the beginning, our new little town had two flourishing taverns—Robert Jones' Palace, at the corner of Main and Third Street South, and the Headquarters Saloon and Restaurant at Shoshone Street and Second Avenue East.

The Headquarters—known to its patrons as the "Bucket of Blood,"—was owned by George Bassett, and leased to S. G. Hamburg. As one man said, the building was "not exactly an architectural beauty, but it had its place in the upbuilding of Twin Falls." Later, sentimental old-timers would recall the days when they "trudged through the dust to the Headquarters for their meals." And their booze.

George Bassett was the first to run afoul of the law. On February 22, 1905, a district judge issued a warrant ordering the officers of the court to search the Headquarters for gambling devices operating in violation of the Idaho penal code, and to seize any such devices found. That evening, Deputy Sheriff F. E. Ramsay executed the warrant with dispatch and took from the saloon's basement a collection of gaming tables, cards and other paraphernalia. The next morning, as ordered by the judge, the impounded materials were piled on a bed of sagebrush and burned in the presence of a large crowd.

Carry Nation
(Courtesy Kansas State Historical Society)

By this time the number of saloons in town had grown to six, and the county authorities decided to take action to rein them in. A state law, passed a few months earlier, prohibited the selling of liquor within five miles of the construction of public works employing more than 25 men. In March, 1905, the same district judge who had ordered the gambling raid, C. B. Taylor, had all six saloon owners arrested for breaking the five-mile law. The judge maintained that the canals then being dug near Twin Falls constituted a public work.

On April 13, 1905, while the bar owners awaited trial, Twin Falls was finally incorporated as a village. Six days later, the saloon owners were tried. They all pled guilty, but, since Twin Falls was

W. L. "Billy" Coltharp's saloon in Hagerman was typical of the era. Here, some off the "boys" are showing off. Perhaps the bartender thinks he is Carry Nation.

The W. C. T. U.

The Women's Christian Temperance Union was founded in 1874, with the goal of eliminating the consumption of liquor.

In June, 1906, the state president of the organization, Adeline Garlock, came with her husband to live in Twin Falls. Within two weeks she had a local chapter up and running with more than 50 members, many of whom were Filer residents.

These ladies destroyed no saloons but, rather, concerned themselves with such peaceful tasks as running a reform school ministry, distributing temperance literature, and lobbying political entities.

School girls often attended the weekly meetings. A Silver Medal contest at Filer, in 1906, was won by an eighth grade student, Elsie Hubbell, for her recitation, "The Drunkard's Blow."

now incorporated, and the five-mile law no longer applied, they were fined only token amounts.

The trustees got right down to the work of passing ordinances and devising ways to raise money with which to run their new village. An ordinance was issued on April 24 that fixed saloon licenses at the price of $2,000 per year. It also prohibited gambling, and required saloons to close between midnight and 7 A. M., and all day on Sunday. Women weren't allowed to enter them after 7 P. M.

The defiant saloon keepers refused to pay the exorbitant license fee, and threatened to lobby for revocation of the village incorporation. Accordingly, all six were arrested for selling liquor without licenses. After one of them was tried on May 20 and found guilty, both sides agreed to a compromise. The trustees would amend the saloon license ordinance so as to enable applicants to pay for their licenses in quarterly installments, instead of paying the entire $2,000 in a lump sum. And, the charges would be dropped if the defendants obtained licenses. The saloon keepers agreed to obtain licenses on the quarterly system, but would test the validity of the saloon ordinance and, in the event they succeeded in

Twin Falls W. C. T.U. destroying liquor in 1922, during Prohibition.

having it set aside, their money would be refunded.

The saloon ordinance was never set aside, but in December 1906, the trustees reduced the cost of licenses to $1,000 per year. Even with this reduction, 37 percent of 1907 village income came from liquor license sales.

The Y. M. C. A.

The Young Men's Christian Association was another group involved in the battle against booze. At the time, the "Y"was a part of the business and social life of nearly every city in North America, providing reading rooms, gymnasiums and a wholesome, Christian place for young men to gather.

The Twin Falls "Y" was formed at a meeting in the Christian Church in late November, 1905.

Using donations, a reading room was established in the basement of a downtown building in January, 1906. With 100 books, it became, in effect, Twin Falls' first library. Various pastors lectured there on subjects such as "The Liquor Business."

Staying in Touch

– Communication Services Reach Twin Falls –

The Portable Post Office

The first Twin Falls "post office" was a small, rough wooden box, nailed to the top of a pole erected at the forks of Blue Lakes Boulevard and the old Blue Lakes to Rock Creek Road. There, the stage driver from Shoshone dropped off the mail for "Cyclone" Bob McCollum, John Hayes and the other earliest pioneers. As the Twin Falls *News* later noted, the letters were generally encouraging. "'The settlers are coming,' said the mails, and the mails told the truth."

When the box on the pole became too small for the job, Harley O. Milner, the prospective postmaster, placed a large valise in the window of Perrine and Burton's general store on Shoshone Avenue. Patrons helped themselves and, if they didn't get the letter they expected, "they had the privilege of kicking the valise or taking a handful of prunes by way of reimbursement."

Milner was formally appointed Twin Falls' first postmaster on August 1, 1904 and seventeen days later he established the post office in a building at 202-264 Third Avenue South, which also was his

Picnic in James Waters' locust grove. L. to r., Mrs. J. A. Waters, Alice McCollum, Mrs. Harley (Lulu) Milner, Mary Milner, Truth Milner, Irene Waters, Zora Waters, Essene Waters, Elva McCollum, Robert McCollum, James A. Waters, and Harley Milner, Twin Falls' first postmaster.

Rosa Roberts drove Twin Falls' first rural mail delivery wagon.

home and lumber yard. Milner installed some crude pigeonholes to serve as post office boxes.

Since Harley was busy operating the yard, and had 46 teams hauling lumber from Shoshone, his wife Lulu, became *de facto* postmistress. "Harley went without his hot cakes many a morning," said the Twin Falls *News*, because Lulu was "too much occupied with the daily mail to attend to the wants of the male." Mrs. Milner would later recall that the men surveying the town were the first to visit the post office in the morning and the last to call at night.

Milner resigned in February, 1905 and was replaced by William W. Dunn, formerly of Hailey. The new man moved the post office to a 13 by 24 foot "cubby hole" in the Tobin Building in the 100 block of Main St. South, which he equipped with tables and a few hundred lock boxes obtained from the Boise post office. But, despite these improvements, the service soon proved inadequate for the burgeoning city, with crowds having to stand in the dusty street for hours, waiting for their mail.

Conditions worsened in August when mail began arriving and departing by rail, and Twin Falls became the distribution center for southwestern Cassia County and southeastern Owyhee County.

Over the next few months the office occupied two other locations before finally—in November 1907—a larger, better-staffed and longer-lasting post office was established in the Babcock building at 121-127 Main Avenue West, adjacent to the Hotel Perrine. In all, Twin Falls' post-office was moved six times during its first 41 months of existence.

The Telephone Habit.

In January of 1906, when the Rocky Mountain Bell Company finally got a decent telephone system operating in Twin Falls, one booster posed these questions: "Have you formed the telephone habit? Do you jump from your chair when the bell rings? Do you forget you have a telephone and walk five blocks to speak to a man who also has an instrument in his office, and then kick yourself? Get the habit of looking at the exchange list before venturing out in the mud. It will save you time, money and shoe leather."

This same fellow thought that the phones being installed in the rooms of the Hotel Perrine would be exceedingly handy for a man to call up his wife and tell her he was "detained by an important meeting of the Commercial Club" and couldn't get home until after the town's lights were turned off at midnight. Thirsty gentlemen could also use the phone to "order life preservers," and ladies could now "gossip to their hearts' content without leaving their apartments." Such luxuries weren't available in any other hotel in the state.

As with most things in early Twin Falls, Burt Perrine got the first telephone system going. The Rocky Mountain Bell Company was already providing phone service to the Wood River Valley and Shoshone when Perrine tried, unsuccessfully, in 1899, to get a line run from Shoshone to Blue Lakes,

Shoshone Falls and Rock Creek. But, in 1903, after work had commenced on Milner dam, the Bell people ran such a line, as well as one to Milner.

"The bustling infant city" of Twin Falls was first connected by phone to the outside world in May 1904, when Perrine ran one line from Blue Lakes. At the time, not a single building had been erected on the town site.

The following month, Perrine and several other men incorporated the Twin Falls Telephone Company, having obtained an operating agreement with Bell. No action was taken until April, 1905, when the company obtained a franchise from the city fathers and began operating a primitive exchange, for a handful of subscribers, out of the transfer company office of J. V. "Sunny Jim" Baker. Baker, who served as the operator, shut the exchange down every evening.

"Sunny Jim" Baker driving Perrine's stagecoach.

The service was slow and unreliable, causing one user to remark that the men seen around town with cauliflower ears weren't prize fighters—they had "merely been trying to get Shoshone on the telephone." Some improvement was made in May when "Sunny Jim" was relieved by two female operators and the exchange began staying open until 11 P. M.

Just as Burt Perrine had probably planned from the beginning, on May 24 Rocky Mountain Bell bought out his company—franchise, pole line and crude plant—for $17,500. Successively more modern switchboards were obtained; new lines were added and, by late October 1905, the telephone office was moved from Baker's office to a building belonging to Perrine's First National Bank, on the east corner of Main and Shoshone.

By January, 1906, as noted above, most of Bell's 58 subscribers were satisfied with the service, including A. E. Bartlett, whose phone number was 13. Within the ensuing year, the telephone office was moved to the First National Bank proper, acquired four operators and began furnishing all-night service to its more than 200 customers.

Telegraph Service

The first telegrams to Twin Falls addressees had to be forwarded by telephone from Shoshone.

The Oregon Short Line Railroad Company completed the first telegraph line into Twin Falls on July 27, 1905, having installed it coincident with the building of the branch railroad to the town.

At first it was operated by a certain C. P. McPherson from William Perrine's farmhouse on the east edge of the original townsite.

George Fraser of the Twin Falls *News* sent first telegraph message from the city.

A Light on the Desert

—The Hotel Perrine—

There was great anticipation when neatly-dressed waitresses entered the dining room and began taking orders for the first meal served at the Hotel Perrine, on the evening it opened. Early on, most of the "Homeless Twenty" men lost their brand new, slippery napkins, being out of the habit of using such refinements.

After the meal, when the finger bowls appeared, Bob McCollum scratched his chin for a moment, then excused himself and made a break for the office telephone to call his friend, Fred Reed, in Burley.

"This stumps me," McCollum told Reed. "Yes, looks something like a mush bowl only it's plain glass. Slice of onion in the water. Yes, might be lemon. Sure, young bed sheet under it. Both hands or one? Just the finger tips. Do you swab off the moustache? So long Fred. Knew you wouldn't see an old pal stuck."

There had been other hotels in Twin Falls prior to the Perrine, but none nearly so elegant. The Twin Falls Investment Company built the stately structure in partial fulfillment of its contract with the Twin Falls Land & Water Company, in which the former had been given five town lots on the condition that it build such a hotel.

Emmanuel Masqueray, the noted French architect, was responsible for

Hazen's Twin Falls Rooming House

The Earliest Hotels

Our town's first hostelry—the Twin Falls Rooming House—was opened in November, 1904 by Scott W. Hazen at what is now 227 Main Avenue East. A second hotel started up the following January, at the corner of Main Avenue East and Third Street East. Called the Burrington, it had just one bathroom for its 20 bedrooms.

Initially, the Burrington's only refinements were "entirely new bedding," and a "ladies' parlor, with a good view of Main Street." Later a dining room, plumbing, and some real baths were added, but the hotel burned down in 1906.

Warren B. Hoag and wife opened the Waverly Hotel, at 125 Third Avenue South, in March 1905. It started with only six rooms to rent, and Mrs. Hoag did the cooking and served all meals, family style, for 25 cents.

Hotel Perrine, at night.

the general concept of the hotel, which he thought should "harmonize...with the topographical and climatic conditions" of the area. Masqueray spent several days in Idaho in late June, 1904, and made sketches presenting his ideas, but the final plans, drawn by architect J. Flood Walker of Boise, contained significant changes.

Excavation began in the fall of 1904 and construction was begun in December under the direction of a local architect, James H. Richardson. The building contractor was Joseph Dedrick of Salt Lake City. The Perrine was initially called Hotel Kimberly, but the name was soon changed to avoid confusion with the new town of that name.

The basement and foundation walls for the three-story structure were made of black lava rock trimmed with red lava, and the building proper was constructed of concrete blocks molded on the site. Its front entrance was set back some distance from Main Street, and was reached through an open court and conservatory, the wings of the building extending on either side to the street. Twelve-foot wide sidewalks were laid around the hotel on Shoshone and Main.

Since Twin Falls did not yet have a sewerage system, waste from the

Twin Falls' first electricity came from the Hotel Perrine power plant.

hotel drained into a huge cesspool dug in back of the building. Also, an electric plant was installed in a shed behind the hotel. It consisted of a threshing machine or tractor gasoline engine, driving a 110-volt direct current generator. Distribution lines were run for a short distance in both directions along Main Street to provide electricity to several other businesses.

The main floor was reserved for the hotel office, a dining room, reception room, barbershop, public baths, a well-stocked bar, and the offices of the Land & Water Company. The upper two floors contained 50 rooms, with private baths in twenty of them. The building boasted steam heat, local and long distance telephones in every suite of rooms, and bathrooms conveniently located in every part of the building. The entire hotel was tastefully appointed with fine furniture and carpets.

With a handsome pennant flying above, the Perrine opened on December 19, 1905. That night, its huge roof-mounted electric sign, containing 300 lights, shot brilliant rays, "like a beacon of progress," far across the desert. The interior of the building was also brilliantly lighted throughout. This first significant illumination on the tract greatly raised the spirits of the people of Twin Falls, who stood in clusters on the street, admiring the scene.

Under the management of Edward B. Williams, and his wife—who previously had started Twin Falls first drug store—the hotel quickly became the center of the city's social life. It was the scene of almost daily card parties, teas, banquets, and such, many of these functions held by the hotel's significant population of permanent residents.

In 1906 a large annex was added at the rear of building, and an elegant 14-passenger, horse-drawn omnibus was put into operation to carry guests to and from the railroad depot. Ira Perrine became sole owner of the hotel in 1909 and it was demolished in about 1968 to make way for a new Bank of Idaho building. The site is presently occupied by Magic Valley Bank.

I. B. Perrine's Twin Falls to Jerome stage about to leave Hotel Perrine.

Little Twin Falls School Building Was Site of First Christmas Celebration

The hardy settlers who were building a new life in the Magic Valley didn't take long to provide for the education of their children. On September 28, 1904, with smoke from the sagebrush fires burning in the town hanging heavy in the air, a committee set out to get donations to finance a Twin Falls school. These men—Samuel T. Hamilton, F. W. Eickhoff and Frank Ramsay—soon raised $600.

In November Ramsay took a census, to justify a request for state funds, and found that there were 80 children of school age within the proposed district, 59 of them in the town itself.

The Twin Falls Investment Company had donated two lots for school sites, but there wasn't yet enough money to build on them. Instead, with the company's permission, a temporary building was erected on a lot at 302 Third Avenue East, where now sits a small church.

S. T. Hamilton

Using a good deal of donated labor and materials, contractor Patrick Hall supervised construction of this wooden frame edifice which, for a time, served as a school, community hall, and as a meeting place for the various church denominations. Within the walls of this little structure, much of the early history of Twin Falls was made.

Even though the building wasn't quite finished, it was used for Twin Falls' first Thanksgiving dance on the evening of November 24. The carpenters had to hustle to get it ready and—to make things snug for the occasion—they boarded up the window openings. A borrowed wood-burning stove and two large oil lamps furnished heat and light for the forty or so couples who danced to the music of the "Twin Falls orchestra," and who took delight in listening to a musical program given by a violinist, Miss Cora Jones of Rock Creek, and a guitarist, "Mexican Joe"—a cowboy from the Shoesole outfit's Point Ranch. The receipts for the evening were $40, which was applied to the school building fund.

It was about the time of Thanksgiving that a few heroic mothers—the Mesdames Ramsay, Taylor, Yates, Griffiths and Hazen—who, since June, had been holding a Sunday school in a tent in Rock Creek Canyon—made plans to have a children's Christmas celebration that would make Twin Falls' first commemoration of Christ's

Albion Normal School

Many teachers got their training at Albion State Normal School, which was created in 1893 by an act of the Idaho legislature.

The settlers of Albion, which was then the Cassia County seat, erected the first building on five acres donated by one of the town's pioneers, Josiah E. Miller. The school opened in the fall of 1896, providing a two year curriculum for teachers. Later, it became a four-year college.

Even in the Magic Valley's earliest days, many of its youngsters, particularly those from Kimberly, Hansen, Rock Creek and Murtaugh, attended the school, which was closed in 1951.

Santa is right up to date in this 1902 model airship.
(Author's collection)

birth a memorable one for every child in the vicinity.

Fifty dollars was raised by donations, and an "ingenious" artist built an eight-foot high "tree" using a large stick as a trunk, with sagebrush branches wired in place on it. The women tied up candy and peanuts in bright paper napkins, and hung them on the tree, along with popcorn, oranges and apples. When the candle holders were snapped on, and red, white and green candles set in place, everyone agreed the tree looked fine.

The Christmas party was held at the temporary schoolhouse on Saturday, December 24, 1904. Some 200 adults and 100 children attended this first large Christian gathering in the town.

A fellow named Charles Tripp, wearing a huge fur overcoat and cap, played the part of Santa Claus, cracking jokes while he distributed the bounty on the tree to the children. He told them about his fine team of 24 jack rabbits, togged out in silver harness and bells, and assured the children that if they had been good they would certainly get more surprises in the morning.

Some of the youngsters gave a program which included the song "Fishermen Are We," sung by Max and Mary Milner, Roy and Willie Grimes, Elva and Clara Tarr and Grace Urie. Mrs. Harley Milner played an organ which had been loaned by Mrs. Ramsay. The $37.10 raised by the party went toward buying an organ for the Union Sunday School.

Twin Falls' first school session opened January 3, 1905—just after the desks arrived—with sixty-three pupils enrolled. The school's principal and sole teacher was H. H. Thornton of Albion, a former Cassia County school superintendent. Thornton worked entirely without textbooks and taught students of all ages in a single class.

During that first winter, the young scholars were kept warm by a small stove, which was stoked with sagebrush from a pile kept at the rear of the school. The children sitting near the stove would be uncomfortably hot, while those at the other end of the room would freeze.

The first school term ended on April 21, with 138 pupils in attendance. Fifty others had submitted enrollment applications, but hadn't been able to be accommodated. By this time, the population of the tract was soaring, and it was expected that 500 students would attend the fall term. Citizens of Twin Falls realized that the little schoolhouse, like the sagebrush, had to give way to better things.

"Magic City on the Sand"

—Making Twin Falls Livable—

With the arrival of warm weather in the spring of 1905, Twin Falls residents were happy to hear that several of the town's stores had ordered handsome soda fountains. It was generally thought, however, that having the fountains would make it "unsafe for a bachelor to walk the streets."

Sprinkling a Twin Falls Street in about 1905.

That spring the citizenry also learned that a new town could be a mighty dusty place. It was so bad, said one observer, that the cheapest way for newcomers to acquire a town lot was to "go out on a windy day and rope one." And—if the wind proved strong enough—it might be possible to "round up a section."

The need for sprinkling streets was one of the reasons that led citizens to begin agitating to incorporate the new hamlet as a village. In mid-March, with the population standing at over 1,000 souls, and town lots selling briskly, a meeting was held in the little schoolhouse building and the idea of incorporation approved.

Subsequently, the Cassia County commissioners were petitioned and, on April 13, they designated Twin Falls as a village. Appointed as trustees (councilmen) were F. W. Eickhoff, a hack line operator; Paul Bickel, a civil engineer; Samuel Hamilton, an attorney; Frank Bradley, a grocer; and Robert McCollum, a real estate salesman. The commissioners also appointed a deputy sheriff and two justices of the peace for the new village.

When the board of trustees organized

The Rink

One of the businesses boasting a soda fountain was "Billy's Place," also known as "Billy's Lunch Room," which was opened as a short order restaurant in February 1905 by William A. Ring. It sat next to the Twin Falls Rooming House on Main Avenue East.

Early on, Ring and a partner bought the cigar business next door, and combined the two buildings to make a 24 by 97 roller skating rink.

The "Rink," or "Ring Theatre," became the most popular gathering place in town, hosting skating parties, dances, public meetings, lectures and concerts. Plays produced there—like "Peck's Bad Boy," "Old Maids' Convention," and "Shall We Forgive Her?"—were particularly popular, as were vaudeville and burlesque shows like the Kempton Komedy Kompany.

Main Street, Twin Falls in March, 1905.

a few days later, Bickel was elected as Twin Falls' first mayor. One of the board's first acts was to designate Albert Snodgrass as city marshal and task him with quelling the all too frequent "reckless shooting" of firearms in town.

In a flurry of activity over the next few days eighteen ordinances were produced. To raise money, license fees were set for a variety of businesses and activities. An annual "road tax" was mandated, requiring every able-bodied citizen between 21 and 50 years of age to either pay $4 or perform two days labor upon the streets.

Gambling was banned, saloons were regulated and "obscene...and indecent" activities were proscribed—such things as using profane language in public places, and keeping a house where "lewd and disorderly persons" assembled together for dancing. It was also forbidden to give or sell tobacco to anyone under age 21.

Other new rules dealt with the "furious or reckless driving" of carriages, leaving horses standing untied in the streets, and the removal of dead animals and refuse.

Our forebears weren't terribly sensitive to the plight of the homeless. While there was some assistance given to indigents—including medical attention—a great animosity existed toward the numerous drifters that inhabited every town in those days, and who were known variously as tramps, hobos or "Weary Willies." In the same year Twin Falls was founded, a mob of Mountain Home men horsewhipped, tarred and feathered, and carried two such drifters out of town on a rail. The councilmen of Twin Falls contented themselves with issuing an ordinance calling for a $100 fine for such "idle or dissolute persons" who were physically able to work, but who had no visible means of living.

Over the next few months, franchises were granted for telephone service and a waterworks. And, with the help of the Twin Falls Land and Water Company, several street sprinklers were obtained to keep the streets settled. Streets were graded—but not paved—and crosswalks of stone or gravel were made. To accommodate the heavy traffic from the south of town, a bridge was built over Rock Creek at the end of Shoshone Street.

The trustees made it unlawful to drive across this "New Bridge," at a gait faster than a walk, or to drive more than ten head of stock over it at a time. A later version of this span would be known as the "Singing Bridge."

How it Became "Magic"

The first time the adjective "Magic" appeared in print to describe our town or area, seems to have been when the Twin Falls *News* dubbed the town "magic city" in its April 20, 1906 edition.

Members of the Eastern Idaho Press Club, meeting here, used the same term in December of that year.

Still later, in August, 1907, a writer for *Outdoor Life* magazine noted that the road from Perrine's ferry connected directly with the main boulevard of Twin Falls, "that Magic City on the sand."

Bridge across Rock Creek looking up Shoshone Street; depot in the background.

Perhaps the greatest challenge for the trustees was in getting an adequate sewer system installed. In the beginning there were only cesspools—pits dug in the ground—to collect sewerage from homes and businesses.

When the trustees proposed a bond issue to build a regular sewer system, there was strong opposition because of the cost. This led the Twin Falls *News* to point out that the odor of the existing cesspools would frighten way more new residents and investors in an instant than would the proposed sewer tax in a decade. The editor posed the question: "Shall Twin Falls be a modern, sanitary city, or must it 'stink and shine...like a rotten mackerel in the moonlight?"

The bond issue ultimately passed, in September 1906, and—after much wrangling over rights of way—the main lines were completed in July of the following year. The effluent was discharged into Rock Creek, where, it was argued, in the rapid stream, it "would scarcely be noticed." This state of affairs continued until 1967, when a modern treatment plant was built.

Another view of Main Street; note board sidewalks.

The Day The River Dried Up

—The Completion of Milner Dam—

On the afternoon of March 2, 1905, with a greatly-reduced stream of water running over Shoshone Falls, a fellow named Harry Wilson—tanked up on "squirrel whiskey"—removed all his clothes, strolled out onto the dry ledge of the falls, and jumped into the river, 210 feet below. Only slightly bruised, he swam to a rock and calmly awaited the arrival of his clothing. Wilson claimed he had jumped simply to show he had the nerve but, in fact, he was trying to draw attention to himself so that he might be hired as a stunt man in a vaudeville show. Later that month, he and his mistress would be arrested and charged with drugging a man in Meridian, and robbing him of $350.

The day before Wilson's daring leap—March 1—the outlet tunnel gates at Milner Dam had finally been closed, allowing water to begin backing up to form a reservoir. Some 2,000 people, mostly from the surrounding towns of Oakley, Twin Falls, Kimama and Shoshone, had been there to witness the notable event, and the fireworks the previous night. A sea of tents had been pitched in the little town of Milner, but accommodations had still been inadequate to handle the throngs of visitors. The noted Utah photographer, Charles Savage—who had taken that famous picture of the joining of the railroads at Promontory Point, Utah in 1869—was there to record the happenings.

Some last minute rip-rapping was completed that morning, and, when all was ready, Stanley Milner, Walter Filer, Mark Murtaugh, Paul Bickel, and other officials, took their places at the gate-closing devices. These men started the gates on their downward course, as Robert Faris and Frank Kesl, "the beavers who built the dam," watched from a nearby knoll. Ira Perrine didn't take part in the ceremony, but stood amongst the crowd gazing "upon the reality of which he had dreamed for years."

Beside him was Robert McCollum, who had "piloted thousands of home seekers over the tract," selling them many acres of water shares and hundreds of town site lots in the previous eight months.

Before long, the officials got tuckered out, and gave way to a muscular gang of workmen. After 45 minutes of cranking, the roar of the water had died away, and the level of the river began creeping up toward the opening of the big south side canal.

But the spectators weren't too interested in watching the water rise. Instead, they

Gates and approach to tunnel under south island.

River drying up on March 1, 1905. Milner Bridge in background.

jostled one another in trying to get a good look at the river bed, as it dried up below the dam, and then hundreds of them went scrambling and leaping across the rocks, heading for the island in midstream. With shouts of glee, they scooped up fish from the little pools and pot holes, and strung them on willows. Also, dozens of men spread out along the river, with cradles and pans, trying to wash some gold dust out of the exposed gravel while the water was low.

One of the gates was obstructed by a piece of rail, which prevented its complete closing, making it necessary to open the gate again, and as the water rose below the dam, people scurried for the banks, and the excitement was over. Most of the spectators then went to luncheon, or started for home.

Frank Buhl and Peter Kimberly missed the ceremony. They had phoned ahead from Kimama, saying they wouldn't be able to reach Milner in time, and asked that the ceremony proceed without them, so that no one would be disappointed. They arrived on the scene during the noon lull, just in time to watch Paul Bickel plant a sack of sand at the foot of the balky gate, to help stop the leak.

Canal Gate Opening Ceremony on March 2, 1905.

The water began to lower at Shoshone Falls, 30 miles below Milner, at about 4 P. M. and, by sundown, the extreme low stage was reached. There had always been disagreement as to how much flowed into the river through the several springs between the dam site and Shoshone Falls. Many thought the falls would be absolutely dry when the gates were closed, but it was found that 2,000 second-feet still flowed over them.

At 10:28 that night, some 12 hours after the tunnel gates were closed, water reached the level of the gates of the south side canal. And, an hour later, the river began to flow over the dam's spillway in a beautiful cataract. On the following day, Thursday, March 2, a ceremonial opening of the canal gates was held. As the gates were raised, Mark Murtaugh's wife, Elizabeth, poured champagne into the first water trickling into the canal.

Within just a few weeks, irrigation water became available to all of the land between the Low Line Canal and the Snake River, east of Cedar Draw, but work would continue on the remainder of the south side canal system for the next several years.

"Provin' Up"

—Early Settlers Did Thirty Days in The Sagebrush—

Rosalia Bohrer, who lived with a family "proving up" near Buhl in 1905, wrote the following to a friend, describing her situation:

> "I am sitting on a bag of alfalfa seed in an unpainted cabin 12 x 14 feet, in the midst of a sun-baked plain covered by low sage brushes, the inhabitants of which are a few jackrabbits, a temporary plover, horned toads, coyotes, an occasional rattlesnake and some of the finest people from various parts of the United States that it has ever been my good fortune to meet...
>
> The country is covered with hundreds of temporary shanties in the sage brush. Gradually the face of the landscape changes and the magic touch of water enables one to look across a country as green as...Wisconsin...and much more productive."

Paul Bickel standing amidst some cute little Idaho "trees."

The Carey Act, under whose provisions the Twin Falls tract came into being, required the settlers to establish residence on their land within six months of being notified that water was ready for delivery. Within a year of such notice, one-sixteenth of the parcel had to be cleared and cultivated. And, within three years, an eighth had to be cultivated, and "final proof" made.

This final step required the entryman to do "thirty days in the sagebrush"—live on the land with his family for a month—and then to publish declarations, in the newspaper, of having met the law's requirements. After this, the state applied to the federal government for the settler's patent to the land.

Proving up was no picnic. At the time, the area was covered with sagebrush, mostly from one to five feet in height. A young Philadelphia woman, while crossing on a train just before the tract opened, is said to have remarked: "I think Idaho is perfectly lovely—the whole country is covered with the cutest little trees...Won't it look grand when they all grow up?"

Unfortunately for the settlers, these "trees" had to be removed, or "grubbed off" before the land could be leveled and a crop planted. Using a grub hoe, a man could clear about half an acre a day. A horse-drawn plow, with its moldboards removed, could clear six to twelve acres. Those with the money could buy a grubbing machine, which cut the sagebrush roots beneath the surface of the soil.

Grubbing Sagebrush on the Twin Falls Tract.

A prove-up tent on recently-grubbed land.

There were also professionals, such as the Johnson Grubber Company, who would do the job for about $4.50 per acre. Ira Perrine and H. L. Hollister owned a contract grubbing rig pulled by a steam traction engine that burned sagebrush to fire its boiler. Their outfit could clear about 40 acres a day.

After grubbing, the brush was raked into long rows about 20 feet apart. Some people kept the fast-burning material for heating and cooking—if there were enough children in the family to keep the fire going. But most set the rows afire, generally in the evening, when the glow of the many blazes could be seen in all directions across the plain.

The entryman's prove-up home was supposed to be a fit habitation for a family, but the rule was not strictly enforced. Usually these "houses" were merely tents, lean-tos or shacks. Only half-jokingly, the story was told of a young bachelor who was afraid to sleep in or near his prove-up hut, for fear it might collapse.

Occasionally the shacks were stolen. And, in a number of cases, they were found to be situated on someone else's property, and had to be dragged to where they belonged. In one instance, a train was stopped by a farmer who, in moving his house, had gotten it stuck on the tracks, despite having three teams tugging at the load.

Living conditions on those first farms were tough. There were rattlesnakes up to five feet long to contend with. One pioneer woman came across one with "about 'steen rattles on his tail," which were singing in a "high minor key." The rattler struck at the lady, but she kept a safe distance, got a revolver and killed the serpent with three shots.

Then there was the ever-present dust from all the land clearing. Though dishes and implements were stored upside down, dirt was said to be on the bill of fare at every meal. It didn't help that water was hard to obtain.

For the first several years, there were no wells on the tract, except a few shallow ones dug along the banks of Rock Creek. Many entrymen had to haul their water in barrels for miles, from the creek or the river. Later, most farm families took their domestic water directly out of canal laterals, filtered it through a trough of fine gravel and stored it in cone-shaped cisterns. A typical cistern was about

Coming To Idaho—"The Colonies"

In most cases, the man of a family arrived on the tract first, built a temporary home, and then went back to his old home for the women and children. Usually the family would arrive with its livestock, household goods, and implements loaded in one or more "emigrant" railroad cars. Before the railroad came to Twin Falls, all the family's possessions then had to be hauled in wagons from the rail terminals at Shoshone or Kimama.

Birds of a feather tended to flock together. The most notable example was in the Filer area, which was originally known as "Sucker Flat" because so many new residents there were from Illinois, the Sucker State.

John H. Crocker, a grain dealer from Maroa, Illinois, was treasurer of the Twin Falls Investment Company, and it was he who first induced others from Illinois to come to Sucker Flat, beginning in September, 1904. Before long, the town of Filer, as well as a school district and small community called Maroa, came into being. Near Buhl, another group from Illinois formed the "Ottawa Colony," named for their old home town.

Many new settlers from northern Idaho congregated on what was dubbed "Coeur d'Alene Ridge," northwest of Filer and just east of Cedar Draw. The principal men involved in gathering together this cluster of families were A. D. McKinlay, and Alfred Page. Many of the northern Idaho men were former miners and, reportedly, one of the first things they did upon arriving was to check for mineral bearing rocks on their lands.

10 feet deep and 8 feet in diameter, and had its interior plastered with sand and cement, or lined with brick or lava rock.

Despite all the hardships, those early settlers were optimistic about the future, as shown by this little verse that appeared in the newspaper, about the time water first came on the tract:

For genial showers you need not wait
You only have to hoist the gate,
And let the waters overflow
Our valleys rich in Idaho.

Moving a Prove-up Shack.

The Busiest Man on The Tract

—Farming, 1905 Style—

Alexander McPherson was surely the busiest man on the Twin Falls tract during its first few years of existence. He was the agricultural whiz—and former state horticultural and pure food inspector—hired by the Twin Falls Land and Water Company to instruct the new occupants of the land in the latest and most profitable methods of farming. That "Alec" won the hearts of his students is evident in this verse from the "Twin Falls Home Song," written in 1905:

"I want to be a farmer and with the farmers stand,
A pitchfork on my shoulder and a shovel in my hand;
I want to join the institute and learn to farm aright,
And go and hear McPherson on every Saturday night,
For he's the canny Scotchman who surely does know how
To cultivate a tater patch or milk a jersey cow."

Nearly every weekend McPherson could be found in one town or another, lecturing in small schoolhouses to one of the farmer's organizations he helped start up. Or, he might be holding one of his "institutes" which dealt with a wide range of farming subjects—soils, land preparation, weed and insect control, irrigation, dairying and fruit growing, just to name a few. In addition, the "superintendent," as he was known, taught basic agriculture to school children and wrote newspaper articles on a variety of farming subjects.

McPherson was also responsible for the "experimental" or demonstration farm which operated for three summers, starting in April, 1905, by agreement between the state and federal governments. Laid out on the east side of Blue Lakes Avenue at its junction with Shoshone Street, the 40-acre farm was used to test various crops and farming methods, using seed provided by the federal government. Similar farms were established for several seasons at Buhl and Milner.

Three years from sagebrush; the McPherson farm, about 1909.
Perrine Coulee is in the foreground.

When McPherson attended an "Irrigation Congress" at Boise in September, 1906, he entered some of the experimental farm produce into competition with that grown in other Western states. The

Picking potatoes, ca. 1908.

Twin Falls potatoes and grains each won a first prize, and the grasses earned a second place. Since McPherson had grown them, he received a silver loving cup, a Pontiac "runabout" wagon, worth $125, and $10 in gold coin. The Twin Falls *News* thought the tall, spare agriculturist would present a striking figure "seated in his swell runabout, with the gold in his pocket."

The superintendent's wife, Caroline, who had been living in Boise, returned to Twin Falls with him to take up residence on the Blue Lakes Avenue farm which McPherson had obtained back in July, 1903 at the first drawing for tract property.

McPherson always advised his neophyte farmers not to work more land than they could handle, and he reckoned that forty acres was about the amount suitable for most families. And, he said, if a family had at least $1,000 available for starting up, they would usually make a go of it. Most entrymen came to realize that McPherson was right about the farmstead size, and many of them who had filed on more acreage, divided it and sold off the excess.

There were about five million of these small family farms in the United States in those days, as compared to less than two million today. On most of them, the work was done entirely by hand or with the help of horses.

It was tough going, especially in the beginning, but the Twin Falls irrigated land held the promise of a good life. A visitor from Illinois summed it up well in a letter to the folks back home, saying: "I doubt if a man in middle life [in Illinois], who is used to lying in bed and listening to the patter of the rain on the roof while his crop is growing, would readily adapt himself to farming by irrigation. But men who have learned the art of putting water on their crops...say that irrigation is the only system which makes results an absolute certainty. Such men never come east to farm."

Hogs, chickens and cows were the mainstay of these small family farms. The woman of the house usually had charge of the chickens, whose eggs brought about 35 cents per dozen. Fattened hogs sold for 5 to 7 cents a pound, live weight.

Unlike today, milking cows was done on a small scale.

Getting Traction

As might be expected, Ira Perrine was the first to bring a "traction engine" (tractor) to Magic Valley.

A huge 30-horsepower steam-propelled machine he and Harry Hollister owned entered Twin Falls on April 19, 1905 in the care of five representatives of the J. I. Case Company.

It arrived pulling a "separator" for threshing grain, a water tank, and two large freight wagons containing plows and coal. It had taken five days for the tractor to drag this heavy load from the railhead at Kimama, crossing the Snake at Starrh's Ferry, with the help of a tug boat.

The shriek of the tractor's steam whistle drew a throng of people wanting to see this ponderous, coal-burning engineering marvel, which was unaffordable to most farmers.

Perrine's rig and the first few others on the tract were used for contract grubbing, plowing and threshing.

Plowing with steam tractor.

Two good Jerseys—the breed generally favored—could provide ample butter and cream for the family table and bring in a bit of cash to boot. In the earliest days, there were no milk processing plants or cheese factories in the valley, and Twin Falls didn't get its first creamery until January 1908, but milk and cream were in great demand amongst the townspeople.

The family cows sometimes made the difference between success and failure. One rustic Scandinavian poet, writing in a 1907 issue of the Minnesota journal, *Farm, Stock and Home*, put it this way:

"But dem feller wit da brindle cow,
Hay got bully t'ing, you bet!
Hay never lose da whole year's crop
If grain ban dry or gits too wet;
Ven hail ban strikin' down da grain,
An' yust ban raisin' fits,
At night hay call dem brindles home
An yust ban pullin ___."

Threshing with steam tractor.

Rabbits and Their Habits

—Rod & Gun Club Brings Respect For Game Laws —

At the beginning of the twentieth century, jackass rabbits—jackrabbits—existed in immense numbers on our Western plains. When the irrigation water began flowing on the Twin Falls tract in the spring of 1905, and the first tender crops emerged on the once-desolate plain, farmers were dismayed to find that the pesky "jacks" were gobbling them up for dinner, and then stripping the tender bark from young fruit trees for desert. Every night the rabbits invaded fields and gardens, and even came, uninvited, into the center of Twin Falls.

Jackrabbit hunters, 1907. L. to right, Frank Lindzy and wife, Frank Benoit and Herbert Martel.

Paul Bickel, chief engineer for the Land and Water Company jokingly suggested that the rabbits might be exterminated by somehow enticing them to eat clay and cement. After filling up on these tidbits, the rabbits would naturally drink from the nearest irrigation ditch, which would cause the cement and clay to solidify and the rabbit would die of indigestion. After the coyotes had stripped the carcass, Bickel theorized, the intestinal formations could be used for anatomical illustration.

But the solution chosen was much more practical. To combat the pests, large organized drives were regularly conducted during those early years, the first being held on May 7, 1905.

Typically, the rabbit "hunters" would assemble in their town, choose several leaders and review the rules of the hunt, which usually allowed only shotguns and clubs to be used as weapons. At the edge of town, the participants would spread out into a line and march toward some cul-de-sac or impediment, such as the canyon rim, driving the rabbits before them until they were trapped and could be dispatched by the club-wielding participants. Those jacks that refused to be driven were shot by armed outriders.

On at least one drive, several of James Baker's transfer company wagons followed behind to carry the carcasses. For what purpose isn't known, but the newspaper warned that it would "not be safe to eat sausage in Twin Falls for weeks to come."

These drives often produced astounding results. One held near Hansen in January, 1906 was said to have eliminated 12,000 rabbits; an outing near Filer, in which 50 horsemen and 300 pedestrians took part, did in 4,000; nearly 3,200 shotgun shells were fired in the course of another.

The drives were, in part, social events. Sometimes, the hunters of one town would challenge those

Ten foot 11 inch, 632 pound sturgeon caught in 1908 at Lower Salmon Falls. C. Lind and George Parton had to rope the fish "like a steer" to land it. A nearby trout fisherman said the men had merely caught one of the "minnows he was using for bait."

of another to a shoot. Frequently, teams within a town competed with one another, the losers buying supper for the winners. After one hunt near Filer, the Ladies Aid Society served lunch for everyone at the Maroa schoolhouse.

Another hunt near Twin Falls concluded with a dinner held at Mrs. William Wolfe's boarding house, at which speeches were delivered on "Rabbits and Their Habits," "The Uncertainties of Rabbit Shooting," and "Wild Animals I Don't Care to Meet."

At this dinner party, preliminary steps were taken to organize a rod and gun club. A few weeks later, in early March, 1906, the club organization was perfected, with the help of Peter Holohan, who had just arrived in Twin Falls from Wallace, Idaho. Holohan, known as the "boy wonder" of trap shooters, was a noted Idaho sportsman and the winner of many trophies.

Property was secured just across Rock Creek, south of the railroad depot. A trap was obtained; a club house built, and one of the member's grandsons was chosen as club mascot. Weekly recreational and competitive trap shoots were held, many of them with other towns.

This may have been the first sportsmen's organization formed in Idaho, and is believed to be the forerunner of today's gun club on the canyon rim.

Although Idaho wildlife was still plentiful in those days, there was little control over the wanton taking of it by game hogs and commercial fishermen and hunters. Near populated areas, deer, sage hen and other upland game bird species had been decimated.

After Idaho had become a territory in 1863, the first legislative assembly passed a law prohibiting the killing of big game during the spring and early summer, but no means of enforcement had been provided. It wasn't until 1903—a year before Twin Falls was founded—that a state Fish and Game Department was established and authorization was granted to hire game wardens in every county.

That was also the year meaningful regulations were promulgated, including a requirement for licences. Women didn't need a license, but resident males, 12 years or older, had to pay a dollar per year to obtain one from a justice of the peace, probate judge or game warden.

For his dollar a resident could kill an elk, two deer, a mountain sheep, and a mountain goat. In season, he was

Magic Valley duck hunter, 1907.

Twin Falls deer hunters, 1907.

allowed each day to take 18 sage hens and grouse, 24 ducks, 3 geese, and 20 pounds of trout, grayling, bass or catfish. And, fishing and hunting seasons were of very liberal lengths. Moose, buffalo, antelope, caribou, swans and most non-game birds were protected.

The Twin Falls Rod and Gun Club was very active in its earliest days in the feeding of quail and other wildlife, and in trying to get game refuges and better enforcement of the game laws. The club's conservation and enforcement efforts were spearheaded by the editor of the Twin Falls *News,* George Fraser, who owned a setter bitch and was an ardent bird hunter. Due in large part to his efforts, in May 1907 the state chief deputy game warden visited Magic Valley and appointed three local deputies. After a few arrests were made, there was a sudden demand for fish and game licenses and a new respect for bag limits.

Fishermen at Milner Dam about 1907.

Amusements Were Simple in Early Magic Valley

On a hot August evening, early in Twin Falls' life, Miss Mamie Horn invited some 30 of her teen-aged friends to her home at the Burrington Hotel, for a well-chaperoned evening of music, games and refreshments. Supper partners were chosen using a method unlike anything imaginable today. Paper hearts, each with a young lady's name written on it, were pinned on a wall, and then the young men were given a toy bow and arrow and asked to hit the heart with the name of the girl they desired to escort. According to report, the marksmanship was terrible, and one poor fellow was so nervous he missed the wall completely.

Dancing around the Maypoles on May Day, about 1912.

Social life and entertainments were simpler in 1904. One's senses weren't constantly assaulted by things noisy, shocking or rude. Wireless telegraphy was just in its infancy, and it would be years before radios were available for the home. There was no television, of course, and motion pictures were just coming to the larger cities.

A few people did own "concert" phonographs, which were brought out on special occasions to entertain friends with musical selections. One fellow near Kimberly invariably enhanced the novelty of his phonograph, by bringing his team of horses close to the porch of his house, and getting the animals to neigh for his guests in response to songs played on the machine.

In those days, people entertained themselves mostly by socializing with one another. There were bowling parties, candy pulls, box socials, "cakewalks," "shadow socials," and "Kensington parties"—whatever the latter might have been. Evenings were spent making music, or in playing popular card games like "high five," "500," "razzle dazzle" and bridge. Also modish were idiotic games like "Squeak," wherein a blindfolded person sits on the lap of another player and tries to guess whose lap it is when the player is asked to squeak like a pig.

Despite the harsh nature of life in a rough, new town, most everyone was involved in some cultural activity, no matter his or her station in life. There were several women's organizations, including the Syringa Club, dedicated to history, literature and art. "Musicales" were regularly held. Even ordinary ladies' parties usually took a cultural or artistic bent. Here are descriptions of two such gatherings held in the same week:

The first—"A literary afternoon was an innovation enjoyed with Mrs. Nixon. An original poem, entitled 'A Flowery Romance,' was given in which the blanks were to be filled with the names of flowers. Mrs. Davis' botanical knowledge won for her a magnificent bouquet of carnations. Light refreshments were daintily served by the hostess."

And—"Mrs. F. E. Ramsay entertained a number of ladies in her home...Tissue paper was given them of which to fashion flowers or animals, and much amusement resulted therefrom."

Children participated in similar refined

This theater, a successor to the Majestic, opened in May 1908 at 325 Shoshone Avenue South in Twin Falls. *(Author's collection)*

The Majestic

Our town's first regular "movie parlor" was opened by Paul Alexander and Carl Tschumy on December 2, 1907. Little Mary Milner, daughter of Harley Milner, won a five dollar prize from the theater owners for suggesting the name "Majestic."

The theater was located over the Mission Cafe at 114-116 Main Avenue North, opposite the Perrine Hotel. An electric sign was put up, which lent a "distinctly metropolitan air" to the town.

With standing room only, that first show started with the comedies "Oh! That Limburger," and "Post No Bills." Then came "My Dear," an "illustrated song" in which the showing of still pictures was accompanied by local vocalists or instrumentalists.

The feature movie that night was "The Bandit King," depicting a gang of desperadoes holding up a stagecoach, and being pursued and brought to justice by a plucky sheriff and his posse.

The Majestic's owners were proud of their modern projection machinery, which was free of flicker and guaranteed to "prevent damage to the eyes." Adults paid a dime, and children a nickel, to see a performance of the program, which was changed three times a week.

Soon it became faddish for the ladies of Twin Falls to get up theater parties, while their hubbies were "at lodge."

amusements. They had their own literary societies, and were expected to recite, sing or play a musical instrument at social gatherings. Little Helen Seebeck celebrated her ninth birthday with a party at which her guests danced around a gaily-decorated Maypole.

As they are today, picnicking, fishing and camping were popular diversions. Clear Lakes, Shoshone Falls and the Blue Lakes were common destinations. One party of four families spent ten days together in Shoshone Basin. Their breakfasts consisted of "trout, more trout, trout fried and fried trout, trout cooked with their heads on and trout cooked without the heads; trout fried with the eyes open and fried with the eyes shut." The group was said to be divided into two classes, "the cooks and the eaters, and the eaters had the best of it."

In wintertime, a good snowfall would invariably bring out a throng of sledding enthusiasts, many of them small boys who installed runners on wooden dry goods boxes. Sledding was equally popular with the beaux of Twin Falls, who, in the winter of 1906, were said to be in great distress owing to the scarcity of sleighs. In Kimberly, some of these eager young men about town induced the blacksmith to fabricate a "cutter," and, in it, a group of young people went to the Thanksgiving dance at Rock Creek, reaching home about 5 A. M. the following morning. Events like these were invariably chaperoned by a married couple.

Dancing was another immensely popular diversion. One of our town's first organizations was the Cassia County Social Club, a dance club for young people, which had its initial gathering in the small temporary school house on December 25, 1904. After the dance that night, as was the usual custom in that era, the dancers repaired to a cafe for a meal.

Later on, dances were held in the upper floor of the Odd Fellows Hall or at Billy Ring's roller

Cast of operetta "The Mikado, " performed in 1912.

skating rink. Several "sagebrush" orchestras, often consisting only of a guitar, violin and cornet, were available to supply the music.

In those days dance floors were given a liberal glazing of wax so that waltzers and two-steppers could glide smoothly and lightly across the floor. Sometimes the waxing was imperfect, as evidenced at one memorable dancing party given at the Land and Water Company office. Burt Perrine was said to have walked across the floor, and made a "very graceful slide which reminded him of the days when as a boy he learned to skate away back in Indiana." But Bob McCollum, unaware of the glassy surface on the floor, tried to cross it and, at the third step, he "lurched to port, lost both propellers and grounded heavily in the keel." It was found necessary to skid him to safety in the rear office.

In another instance—at Buhl's first Thanksgiving dance given in 1906 by the Odd Fellows Lodge—the floor got covered with a thick layer of paraffin that "proved better adapted to catching flies than to making dancing easy." The two step became "any old kind of step to get the feet loose from the floor" and, at the evening's end, folks thanked the arrangements committee for giving them the opportunity to "exercise their limbs and take lessons in high stepping."

Picnicking and fishing on the river about 1906.

Matters of Life and Death

– Staving Off the Grim Reaper at The Turn of The Century –

Dr. Henry W. Clouchek

Four-year old Bertha Spencer died of measles on February 16, 1905 in her family's tent home, in Rock Creek canyon. Her 23-year old mother, Gertrude, died a week later of the same disease. Although F. W. Havlicek had recently opened the town's first undertaking and embalming parlor, Mrs. Warren Hoag—who with her husband ran the Waverley Hotel—went to the Spencer tent and prepared Gertrude for burial. She also did the washing, baked the bread and took care of the surviving children while Mr. Hoag took the bereaved husband in a wagon to Shoshone to buy a coffin.

Death was always close at hand in the earliest days of Twin Falls' existence. If measles didn't strike a person down a multitude of other killers—smallpox, pneumonia, tuberculosis, typhus, typhoid fever, influenza, scarlet fever, and more—were waiting in the wings to do the job. Life expectancy was 46 years for men and 48 for women, as opposed to 74 and 79 today.

Newborns and young children were the most vulnerable. One of the earliest to perish in Twin Falls was the two-month old child of the George Battys, who died of pneumonia a month before death took the two Spencers. Batty was a freighter plying between Shoshone and Twin Falls, and the couple also lived in a Rock Creek tent.

Of the 141 deaths during our community's first three and a half years, 29 of the victims were infants and 17 were young children, representing, together, 37 percent of the total. Also five mothers died from the complications of childbirth which, in those days, nearly always took place at home.

Two young doctors, Henry W. Clouchek. and Walter F. Pike, both of whom arrived in October 1904, were the first to take up medical practice in Twin Falls. Several months later, Dr. Truman Boyd hung up his shingle, as did Dr. Jane Shank, an osteopath. They all soon had their hands full in the growing town. These were the days when doctors often actually came to the patient's home, and thought nothing of traveling miles over rough roads, even at night, to render treatment.

Kitchen Table Operations

Before Twin Falls had a hospital, serious operations and complicated cases were sent—if possible—to hospitals in Salt Lake City, Pocatello or Boise.

Otherwise, the procedures took place in the doctor's office, or in the home of the doctor or patient. They were generally performed with another doctor assisting, perhaps because there were so few trained nurses.

In one instance a serious case of appendicitis developed requiring immediate treatment. The patient could not be moved to a distant hospital so Dr. Pike fitted up a room in his building to which the patient was taken. There, Boyd and Pike performed the operation while Dr. John B. Coburn administered the anesthetic (chloroform or ether).

The First Hospitals. Dr. Boyd was running his

Twin Falls' First Hospital, ca. 1907. Expanded and remodeled several times it eventually became the Park Hotel.

practice out of a small building in the 200 block of 2nd Avenue South in the fall of 1905 when a serious outbreak of typhoid fever occurred in southern Idaho. Initially, Boyd pitched a large tent behind the building for his typhoid patients. Then, a short time later he leased a small shack, one block to the east, and turned it into an 8-bed hospital.

In mid-October, Boyd convinced the Twin Falls Commercial Club to get involved in organizing a real hospital. About $1,745 was collected from various people, but this had to be used to maintain Boyd's temporary hospital which, by then, contained a half dozen typhoid patients. Businessmen also donated materials, such as coal and utensils, and the ladies of the town prepared food for the patients.

The hiring of two trained nurses—the Misses Voltz and Weinhold—led a number of young men to feign sickness in order that they might also be nursed. The guilty parties were warned that they would be ducked in Rock Creek and made to drink its water if they didn't straighten up.

The Commercial Club appointed a permanent hospital committee that winter, but there was no further activity until May, 1906 when Drs. Pike and Boyd, with several citizens, incorporated the Twin Falls Hospital Association. By July, the association had subscribed $4,050, and the Townsite Company had donated eight lots in the 200 block of Third Avenue West (which is now the loading area and rear entrance for the downtown post office).

In October, coincident with the onset of another, more severe typhoid epidemic, construction of the hospital was begun. Work had to be suspended for about six months due to a lack of money, but

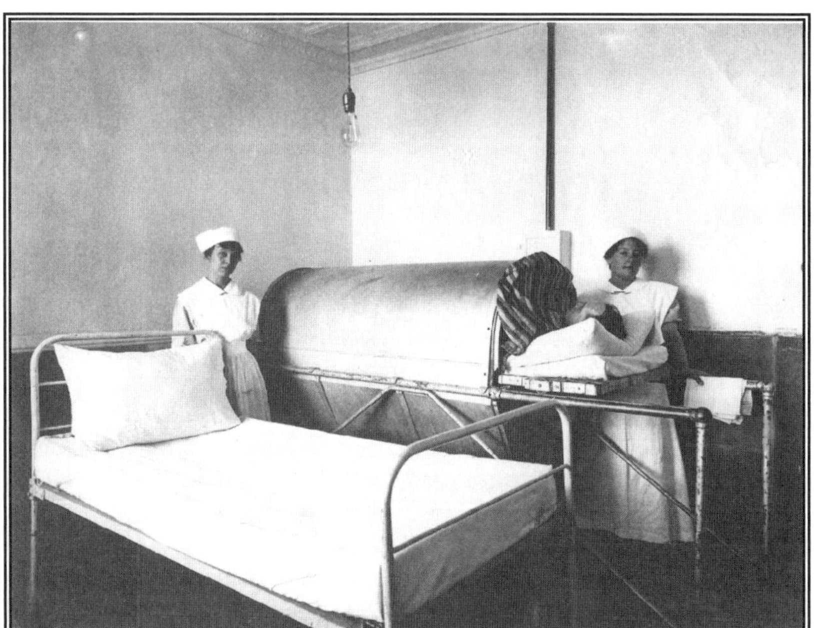

Nurses in T. Falls Hospital, ca. 1910. The patient may be in a hot air machine, described by Dr. James Monroe of Buhl as being designed for "cooking the rheumatism out of the human system." *(Courtesy T. Falls County Historical Museum)*

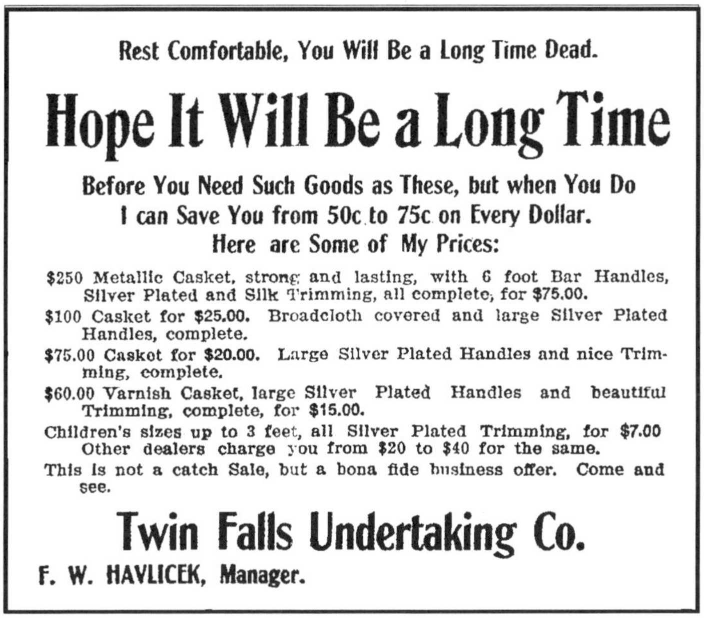

Touchy-Feely Ad run by T. Falls' First Undertaker, in December 1906 *(Author's Collection)*

more was raised by holding a series of raffles, bake sales, plays, dances and a "fat man's baseball game." By December of 1907, the building was completed and fully operational, staffed by a matron and competent nurses.

The Old Burying Ground. The first recorded death in Twin Falls was that of Mrs. Roy W. Gager, the 26-year old wife of a Twin Falls hardware merchant, who died of heart failure in late October, 1904. The Rev. H. W. Parker, a Methodist minister of Shoshone, presided over the interment in ground belonging to a civil engineer named Herbert Martin. The land lay about a mile or so from the center of town, and Martin had apparently indicated he would donate it for use as a cemetery.

Mrs. Maud Powell, lessee of the Burrington Hotel, died of a stroke while drying dishes, on May 9, 1905. Her burial, two days later, was the first at which there was a funeral procession of carriages. The road to the cemetery was circuitous and almost impassible since there were no bridges over the irrigation laterals and ditches. At one place, the pallbearers had to remove the coffin from the wagon and carry it over a ditch.

Mrs. Powell's death drew attention to the need for a maintained cemetery and a better road to get to it. At the organizational meeting of the Twin Falls Commercial Club, held July 11, committees were formed to obtain rights of way for the road and to get a deed from Herman Martin for the cemetery. But, by this time, Martin had decided not to donate the valuable 20-acre tract to the city.

The problem festered for about a year, unresolved, as more bodies were buried on Martin's property. Finally, in August, 1906, after strong editorials by the *News,* a Twin Falls Cemetery Association was incorporated by a recently-arrived undertaker, Haskell Hunt, C. Harder and others. They had the site that exists today on Kimberly Road surveyed and platted.

In the spring of 1907 Martin leased his "Old Ground" cemetery land to a farmer. At least 28 people had been interred on the property, and the lessee gave notice that they must be removed by May 1. Records show that undertaker Hunt, moved 24 bodies to the new cemetery, at a cost of $23 per corpse. But, as of late November, the last time the subject was mentioned in the paper, there were still a number of bodies to be removed, most of them indigents, or "Italians killed on the [irrigation] works," who had been buried at county expense.

Residents of Twin Falls, who live between Sherry Drive, on the north, and Elizabeth Street on the south, and between Sunrise on the west and Eastland on the east, may want to be careful this spring about how deeply they dig on their property.

Bickel

—Twin Falls' New Brick School—

Twin Falls' first permanent school building was dedicated on August 24, 1906. After prayers were offered by two local clergymen, brief speeches were delivered by the mayor and Miss Belle Chamberlain, assistant state superintendent of public instruction. Belle owned a farm on the tract and was in town looking over her corn crop.

Next, the Twin Falls municipal band performed, after which the multi-talented music teacher, Miss Ethel Manker, sang in a duet, played an instrumental solo, and recited "The Reason Why My Feet Are Big." But the most popular entertainment was provided by three groups of little girls. One bunch executed a "milkmaids' drill"—whatever that might have been—and another group did something called "Three Kitty Cats' Night Out." The third batch portrayed "Six Modest Quakeresses."

Twin Falls' initial school session had only run from January 3 to April, 21 1905, and had been held in a small wooden building that also served as a community center. At the end of this first term, Superintendent H. Thornton asked for construction of a brick building with at least eight classrooms, and for institution of at least two years of high school in the fall.

So, with a progressive spirit, on June 5, 1905, the voters of Twin Falls approved the tax levy necessary to provide a decent school system. Cassia County District 27 was established and C. D. Thomas, A. N. Sprague, and E. B. Williams were elected as its trustees. An architect drew up plans for a brick school, and in late June, the citizenry unanimously approved the issuance of bonds to build it. Two blocks had been aside by the Twin Falls Townsite Company for schools, and, at a public meeting held on July 3, one of them, Block 77, was chosen as the site for the new school.

The trustees authorized a salary of $1,000 per year for a principal and asked for bids on 400 student desks, three globes, 10 teachers' desks, four world map sets, eight dictionaries, two reading charts, and one chart of human body. A list of textbooks was prescribed which included titles like "Sunbonnet Babies," "Overall Boys," "Heath's Primer," and "Jones' Reader."

Due to delays in getting bids, a building contract for the brick building was not signed until mid August. Since it could not be completed in time for the fall term, it was necessary to award contracts to erect, on Block 71, two temporary 24 by 72 foot wooden buildings, to be ready by September 18. But,

The Teachers

Twin Falls' early teachers were very active in community and church affairs, and in the cultural life of new little town. Many were accomplished musicians and dramatists. Some of them bought town lots or "proved up" on parcels of farm land.

Mostly female, young and unmarried, they were also heavily involved in the town's social life. They went sledging, dancing, and on moonlight wagon rides with the many eligible bachelors.

Several of the women were active in temperance organizations, and some headed committees for the county fairs.

For their salary of about $70 per month, they not only taught, but supervised a wide variety of extracurricular activities—sports, debate and speech contests, societies and clubs, and parties.

Paul S. A. Bickel

because the desks didn't arrive on time, classes for all of the 235 enrolled children didn't begin until October 16, when finally everyone could be accommodated in one or the other of the three temporary buildings. Four teachers and Thomas Potter, the new superintendent, comprised the school staff at the beginning of the session.

That October, 1905 day marked the beginning of the first central rural school system in Idaho. Despite primitive roads, the district trustees had provided horse-drawn wagons as busses to bring in children, free of charge, from adjacent farms—even those outside the district, who, as yet, had no schools of their own. Two rural routes had been established, running along section lines.

Soon, two new, 25-passenger Studebaker wagons arrived, which were equipped with side and front curtains, and were heated by means of porcelain bricks. In the afternoons, when school was out, the pupils were allowed to pick up their parent's mail at the post office, or do what shopping had been entrusted to them, before they scrambled on the wagons for their homeward trip. Some kids used the delay to dig for fresh water mussels from nearby laterals.

During the summer of 1906 the brick school was finally completed, at a cost of $29,102. It was named "Bickel" in honor of Paul S. A. Bickel, the chief engineer of the Twin Falls Land and Water Company, and the village's first mayor. Built of "Salt Lake compressed brick," the structure was heated by steam and contained twelve 70-pupil classrooms, two lavatories, a principal's office and a 54 by 83 foot assembly room. A large bell was hung for use as a fire alarm and time indicator.

Some 395 students began Twin Falls' first full year of school in the new building, on September 4, 1906. Thomas Potter had been re-hired as superintendent, and S. D. Parke, formerly at Albion Normal, was named principal. Before the year was over the school was running seven rural bus routes, and thirteen teachers were instructing over 500 students.

The forty or so students in the high school studied subjects long since discarded by educators—Latin, "Cicero," and "Caesar," for example. They could also take bookkeeping, shorthand and typing. In addition, there were scientific, domestic and manual training courses.

Bickel School under construction, 1906.

Students in the fourth grade and above were offered agriculture classes, which featured a popular weekly lecture by the Land and Water Company's expert on the subject. A small plot of ground, for use as a miniature "experimental farm" had been donated by Paul Bickel, who lived nearby.

There was little effort made in those days to soothe the "self esteem" of those students doing poorly. The newspaper routinely listed honor students, high school dropouts, and the rankings on the exam for high school. Subjects discussed at teacher's meetings included "How to stimulate interest and activity in dull pupils," and "Shirks in school."

The Bickel School of today was constructed in 1938. The original school was to have been refurbished and left in service but, due to severe water damage during a fire, it was torn down.

Bickel School with wagons lined up loading students, ca. 1906.

The Hose Cart Brigade

—Early Twin Falls Fire Fighters—

The cynical editor of the Twin Falls *News* was upset and frustrated by the failure of the latest version of a fire department to have extinguished a single blaze. "Some months ago," he wrote in February 1907, "the city bought two hose carts and hose, and it was announced that the city had fire protection. These carts were run out on Main Street, the gongs were clanged a few times to propitiate the fire gods and, since then, nothing has been done...No doubt...when a fire breaks out [the fire carts] will automatically sound an alarm, speed to the scene and, couple their hose to the hydrants and extinguish the fire without the...touch of human hands"

As with many other civic improvements in early Twin Falls, the impetus for fire protection originated with the Commercial Club (Chamber of Commerce). The subject was first brought up in September, 1905, a little over a year after the town was begun. In November, the People's Cash Grocery was destroyed in the village's first conflagration.

Formation of a volunteer fire brigade was talked about over the succeeding months, but nothing was done, save the promulgation of a fire ordinance by the village trustees. Finally, in May 1906, a young surveyor, Hugh Austermell—with help from the Commercial Club—put together a loosely-formed organization of about forty men.

But these volunteers had no equipment, except some old hose, and, although four hydrants had been installed, the city water pressure was inadequate for firefighting. At the time, a gasoline-powered engine was being used to pump water from wells dug in Rock Creek canyon to an elevated wooden tank, whence it was distributed to the built-up areas of town. Thus, when the Burrington Hotel burned to the ground one August evening, the water pressure from a nearby hydrant lasted only long enough to wet down and save the tent of a minstrel company performing next door, and the Twin Falls Rooming House and Ring Theater across the street.

Twin Falls' original city water tank (left) was replaced with a new one (right) in mid-1906.

In October, the Townsite Company gave two lots to the city, situated where The Ballroom (formerly the Elks Lodge) stands today. Two temporary school buildings

Volunteer firemen with hose cart, in front of temporary fire house, ca. 1907.

were then on the lots, and had recently been used for the first Twin Falls County fair. One of these buildings was purchased for use as the fire station.

When two hose carts and 1,000 feet of new hose arrived, they were placed in the front of the building, facing what is now 2nd Avenue North. For a fire alarm, a huge triangle — which made a noise that would "put a coyote to shame" — was hung in front of a nearby cigar store.

Meanwhile, the first volunteer brigade having broken up, another company had been formed. Several trial runs and surprise tests of the new hose carts were reportedly conducted, without alerting the Waterworks Company, and pressure was normal. About November 27, a wooden pipeline from a new pumping plant and reservoir on the low line canal was completed, and connected to the city system and its new steel storage tank. Henceforth, as was claimed, there would be 90 pounds of pressure available to serve the 13 hydrants which had been installed by then. The village — the Twin Falls *News* trumpeted — was now protected against fire.

But the temporary fire hall collapsed during a fierce December windstorm and, in January, 1907, another operational test failed, due to a leak in the city water main. These latest incidents were what prompted publication of the sarcastic editorial cited above. This, in turn, led the City Council to get more involved. A budget of $400 was authorized and, on February 20, the brigade was reorganized with Frank Lindzy — a furniture store owner — as chief. Lindzy was to be paid $15 per month, and the other twenty volunteers would get $1 for each fire they attended. An award of $5 would be given to the first company reaching any fire.

As a consequence, speed records were broken at the next fire drill. The blowing of the whistle at the electric light plant in the rear of Hotel Perrine—agreed upon as a fire alarm—caused a hurried stampede for the fire station and, in just under four minutes, hose cart No. 2 was standing in front of Harder's grocery at 117 Main Avenue East. As the newspaper reported, "the boys thought that Mr. Harder, being a member of the board of trustees, would hand them over the $5 prize for being the first on the scene, but they didn't have breath enough to ask him..."

But one test does not a fire department make. Continuing troubles led to another reorganization in late March, and still another when the chief resigned in late April. A bad fire occurred, on June 30, 1907, destroying a new three-story building and an adjacent pool hall. The fire department arrived with a cart within five minutes, but someone had inadvertently closed the valves at the city water tank.

Small improvements were made over the next few years—more hydrants were installed, a fireman was stationed in the hall at night; a telephone and a fire bell were provided. But real progress would not be made until 1909, when a $10,000 bond issue was approved for fire department improvements.

In succeeding months a two-story engine house was built at 236 Second Avenue South, two full-time firefighters were hired, and up-to-date fire engines were procured. Thus was built the foundation for today's excellent department.

Steam fire engine, purchased in 1911. The horses were named Tom, Dick and Harry.

The "Irrigators"

—Baseball in the Sagebrush—

It was a time when baseball was truly America's game. A time when men of all ages and occupations loved to take off their coats and ties, roll up their shirt sleeves and join in the fun.

The major-league "boys of summer" at the turn of the twentieth century included such legendary figures as John McGraw and Denton T. "Cyclone" Young, of the Boston Pilgrims. It was Young who, on May 5, 1904—about the time the town of Twin Falls was being surveyed—pitched the first perfect major-league game against Connie Mack's Philadelphia Athletics, during a string of 44 consecutive scoreless innings.

During the building of the Milner dam, energetic baseball games had been played by the young engineers and functionaries in the area, but no team was formally organized in Twin Falls until about the time water was first turned onto the land.

Readers of this column may remember John E. Hayes as the civil engineer who laid out the central part of Twin Falls, with its 'cockeyed" streets. In early March, 1905, Hayes hung a sign-up sheet in the window of the Liberal Cigar Store—which he owned with A. N. Frankel, and which was on Main Avenue South—and began organizing a city team. The 28-year old Hayes had caught a bit for the St. Louis Browns when he was younger. Those were the days, he would later tell his grandson, when catchers stood several bounces back of home plate and didn't move up behind the batter until after two strikes had been called.

Twin Falls cigar store, probably the "Liberal." Note the liberal number of spittoons.

Toward the end of March the Twin Falls "Irrigators" came into being, with Hayes as manager and Lou E. Kiersted, who worked at the Liberal, as captain. At first, there was no decent playing field, so practice was held in the street, in front of the cigar store. But, in early May, Paul Bickel volunteered a piece of his land at the southeast end of Main, near what is today East Five Points. The players cleared this parcel, laid out a diamond and—to raise money for equipment— they held a dance in the new Odd Fellows hall, at which ladies escorted by gentlemen were admitted free.

A good deal of hype in the

newspaper preceded the actual playing of any games. Lou "Christy Mathewson" Kiersted, the Twin Falls pitcher, was said to have a repertoire of mystifying pitches that included the "Shoshone Falls drop," and the "Auger Falls twister." He had also perfected the "fit ball," which curved like the inverted siphon that carried the Lowline Canal under Rock Creek, causing a batter to throw a fit when he saw it rise suddenly in his direction.

Hayes, the catcher, had made two doubles, a "triplet" and a home run in a recent game in Burley, so, for the Irrigators' forthcoming contests, several small boys, allegedly, would be posted in Will Perrine's pasture, next to Bickel's place, to retrieve balls when Johnny was at bat.

In a similar vein, Charles Mull, a civil engineer, said he had found himself an old wagon tongue to use as a bat, and promised he would "lam the ball into the low line canal every time up."

The Irrigators' first game was played against the team from Albion, the county seat, on Saturday afternoon, May 27, 1905 at "Bickel's potato patch." Hayes returned from Burley—where he was surveying that new town site—just in time to play. The team had invited "the handsome women of Twin Falls"—every lady in the city—to attend and, for the first time, cheer for the home town at an athletic contest. Patrons of the game were asked to please stay on the road to the ball grounds and not tramp through Bickel's budding orchard.

The ground was loose and dusty at the beginning of the game, but a heavy rain shower laid the dust beautifully, while also putting crimps in a few of the ladies' spring hats. After all the pre-game buildup, Lou Kiersted didn't pitch for Twin Falls. Instead, Carl Camp, who pitched from the "Cedar Draw [left] side of his anatomy, took the mound. What with his blue jeans bound round his shoe tops with string, and his wide-brimmed straw hat "set rakishly in the direction of Rock Creek," Camp was said to look like "Uncle Josh Whitcomb"—a hayseed farmer in a then-popular play called "The Old Homestead."

Johnny Hayes, wearing the mask and chest protector, was said to have stopped "many a wide pitch gone awry." Another standout was hardware store owner, R. C. Snyder, who "covered about four acres in the vicinity of second base."

Early baseball game held on Bickel ranch.

The Bruins and Cowboys

Twin Falls has been home to several minor league, professional baseball teams. The first was a club called the "Bruins" that played in the Utah-Idaho League from 1926 until 1929, when the league was dissolved.

The team's first manager was Carl Zamloch, a pitcher for the Detroit Tigers in 1913 and 14, and a teammate of Ty Cobb. The club was at first called the "Magicians" after Zamloch, a practitioner of that art.

The Twin Falls "Cowboys" played in the Pioneer League from its inception in 1939 through 1971. Home games were played on a field at Harmon Park, which had a grandstand, night lighting and other amenities.

During its lifetime, the club was a farm team for Seattle, of the Pacific Coast League, and also for the New York Yankees.

The Irrigators narrowly defeated Albion in the ninth inning, 16-15. That evening, the Albion players, whose gentlemanly demeanor "won the respect and good will of the people of Twin Falls," were the guests of the Twin Falls club at a dance in the Odd Fellows hall.

More games were played, that year and the next, against the Burley "Ditch Diggers," the Filer "Rattlers" and other teams from nearby towns. The home games at Bickel Field were occasionally interrupted by jackrabbits trespassing on the diamond. When this happened the game was temporarily forgotten while players assaulted the interloper with bats, balls and clods of dirt.

Twin Falls Bruins, ca. 1927. Manager W. W. "Frenchy" Frantz is seated far left, first row. *(Courtesy Arthur Frantz of Twin Falls)*

The Park

Residents of Twin Falls are justifiably proud of their fine central city park. A newspaperman in 1907 correctly predicted that in years to come it would always be "a beauty spot, a playground, and a breathing place." The second part of his prediction, however—that "no man who put up a fine house facing the park would have a jail stuck up in front of his door"—was not so foresighted.

Four blocks were tentatively chosen for a park on July 5, 1904, during a visit to the still-barren townsite by Frank Buhl, Peter Kimberly, and others. Two of the blocks were on each side of Shoshone Avenue, where the park, courthouse and jail are now situated. At this meeting, plans were made to plant shade trees on an extensive scale, both in the park and throughout the town.

The man responsible for getting things started was Alexander McPherson. In early 1905, the Twin Falls Land & Water Company hired McPherson, a competent agriculturist, to instruct and assist the many new farmers on the tract.

In April, McPherson and Chief Engineer Bickel of the Land and Water Company inspected the proposed park grounds and decided how to lay things out. Grass would be sown on the entire plot, and tree roses planted here and there to adorn the walkways. A row of elms would be set out all the way around on the perimeter. Inside the sidewalk line, there would be a row of other tree varieties. The entire parcel would be curbed and a four-foot strip of lawn planted between the curb and sidewalk. The first irrigation water had recently been turned onto the tract, and ditches would be dug

Park and town, ca. 1910.

Young Trees along Shoshone Ave. Park is in foreground; Hotel Perrine in distance, at right.

with which to water the trees and lawn.

McPherson obtained several thousand saplings from a Boise nursery—silver leaf maples, ash, black walnut and several varieties of elm—which were planted that spring in the park, and also along Shoshone Avenue and Blue Lakes Boulevard, from the south side of town to the top of the Snake River canyon grade.

Another of McPherson's responsibilities was running a forty acre "experimental" or demonstration farm, located on the east side of Blue Lakes Avenue at the north five points. Crops of all kinds were grown there and different methods of watering tested. One of the crops planted on that farm in 1905 was Central Park lawn grass, which would be used the following year to reseed the poorly-growing park lawn.

By 1906, it had been decided that part of the four blocks originally set aside for a central park would be used for public buildings—a high school, city offices and a Carnegie library. To make up for the partial loss of park land, there was talk, for a time, of transforming 15 acres of Rock Creek Canyon into a city park. This proposed park would have run from the city's waterworks well to a new bridge that was soon to be built at the foot of Shoshone Street. It would have been three quarters of a mile

Buhl Sandnes Got His Bath

Frank Buhl was known as a generous man. He not only gave the city the lots for a park, but also for a hospital, churches, schools and county buildings. He was also always willing to pony up cash for projects like street sprinklers, bridges and the county fair.

In the fall of 1906 he visited his namesake town, Buhl, and, while there was told that the first baby boy born in that town had been christened "Buhl" by his parents, the John C. Sandneses. Frank Buhl promptly made the boy a gift of two town lots, and opened an account for him in the new bank, of which Buhl was a part owner.

When Buhl asked if the waterworks system ran as far as the Sandnes residence, he was told that it did not, and that the cost of extending it would be $2,500. "That's all right," said Buhl: "make the extension; that boy must have his bath."

long, complete with roads, walkways and bridges across the stream, and the entire parcel would have been planted to trees.

These plans were evidently dropped within a matter of months when it was decided to build a city sewer system that would empty into Rock Creek, not too far distant from the site of the park.

In July, 1907, after beautifying and maintaining the central park for two years, Frank Buhl, on behalf of the Twin Falls Town Site Company, formally turned over the two blocks on which it is located to the city.

The park has changed somewhat over the years. The lovely wooden bandstand was replaced during the depression era with the present shell of concrete and lava rock; a decorative fountain has been installed; diseased trees have been replaced. But, despite these changes and ever-increasing use, the city park still remains a special island for repose and contemplation in our growing city, and is a wonderful place to spend a summer's evening.

Park in its Prime, ca. 1918.

And The Band Plays On

Twin Falls most enduring tradition is its city band. Most small communities had such groups around the turn of the twentieth century, when our town came into being. They were frequently called "Silver Cornet" bands and, in those days before radio and television, their rich and mellow sounds on a warm summer evening were a delight to our forebears' ears.

The Twin Falls "brass band" was organized in April, 1905, about the time the gates were first closed at Milner Dam, and water began to flow across the Twin Falls tract. One of the few places to meet in early Twin Falls was William A. "Billy" Ring's roller skating rink, and that is where the musicians gathered to chose officers.

Named president was Charles Thomas, the local state land agent. George Fraser, editor of the Twin Falls *News* was chosen vice-president; Charles Cole, who ran a clothing store, became the secretary; and, J. M. Maxwell, a bank cashier, was picked as treasurer.

Although it was announced that instruments would be purchased immediately, there was no further activity reported for over a year. Then, in April 1906, the local Odd Fellows celebrated the 87th anniversary of their national lodge. It was announced that the Twin Falls Brass Band would play for the occasion but, for some reason, the Silver Cornet Band from nearby Shoshone played instead. As the *News* reported, this was the second band to "awake the echoes of Rock Creek Canyon." The first had been the Pocatello City Band, which had come to town for "Railroad Day," in August 1905.

Shoshone Silver Cornet Band playing on Main St. for Odd Fellows celebration, Apr. 26, 1906.

Twin Falls Band in front of original bandstand, ca. 1910.

This incursion by the Shoshone band seems to have provided the necessary stimulus to the Twin Falls musicians because, a few weeks later, on May 16, a number of them met and decided get the band re-organized immediately. Among those present were George Lacy (a farmer and clarinetist), John Bainbridge (a real estate agent and clarinetist who had once played with the Northwest Mounted Police band), Carl Camp (cornetist and owner of a cement company), and his father, Silas B. Camp.

Twin Falls band leads 1910 county fair parade.

The elder Camp — an accomplished musician, who had just arrived in town with his cornet — became the first director.

Some members already owned their instruments, but more were needed for those who didn't. Camp rehearsed those who were equipped, and the group made its initial public appearance at the Independence Day celebration on July 4. With faint praise, the *News* reported that the music had been "very acceptable," and assured readers that when all the instruments arrived, and the bandsmen had "a little more time to practice," the band would be "one of the best in the state."

More instruments arrived, and a much-improved ensemble gave its very first concert in the park, on the evening of August 10. A modest bandstand was erected, which was lighted with electricity supplied from the Hotel Perrine's gasoline-powered generator, a few blocks away. It would be another year before the Shoshone Falls power plant came on line to totally electrify Twin Falls.

The program for this first park concert included the two-steps "Battleship Maine" and "Uncle Sammy;" the polka "Oneoto; a schottische, "Darling Vincent," and the Symphia waltzes by Holzmann. The band also performed August 24 at the opening of Twin Falls' first permanent school building, named in honor of Paul Bickel.

Ad for Band Minstrel Show held in April, 1914. A reviewer thought it had been "rendered...with a precision and snap that kept the audience well-pleased." The "end-men"and "interlocutor," were particular favorites.

In other appearances that first summer and fall, the band performed in concert at the skating rink, played for the opening of the Kimberly railroad depot, and provided music from a specially-built bandstand at our first county agricultural fair, held in downtown Twin Falls. In November, the group played for a Democratic Party rally, and on New Year's Eve, gave a musical program that included a cornet solo by Silas Camp.

In the summer of 1907, on Thursday evenings, the band gave five open-air concerts to large crowds in front of the skating rink, and then played for dancing inside. Also, that year, the group played for occasions as varied as Memorial Day services, the fair, the opening of Sweet and Gaut's saloon, and the reception for William Jennings Bryan, who was visiting Ira Perrine.

A high, round bandstand was built in the city park, using band funds in 1910, and weekly summer concerts began there on Friday nights, soon afterward. The present semi-circular band shell was built on the same site during the 1930s, with the help of the Works Project Administration, a make-work organization that existed during those years of economic depression.

It can't be determined with any certainty who the earliest directors were, subsequent to Silas Camp. In 1908, it was reported that the band would reorganize under George Reels, an experienced bandmaster. But, the following year, a certain Frank H. Thomas was mentioned as having been director. Another source has C. D. Tyler taking over from Camp; still other documents have John Bainbridge as director from 1907 to 1922.

Later Directors of the Band

- Jack. C. Thorpe: 1922 - 1926
- Dr. Orrin Fuller: 1926 - 1944 and 1946 - 1958
- Charles Ratcliffe: 1944 – 1946 and 1958–1967
- Del Slaughter: 1967 – 1978
- Ted Hadley: 1978 to present

Railroad Day

—The Building of the Minidoka and Southwestern—

Not long after the railroad reached Twin Falls, a passenger coach on the "lightning express" caught fire while standing in front of the depot. The sarcastic editor of the local newspaper thought the fire, which started in the roof, might have been caused by "friction resulting from high speed." He hoped that the loss would not be cause for the Union Pacific tycoon, Edward H. Harriman, to increase passenger rates.

These comments typify the love-hate attitude most area residents had toward the railroad. While they were happy that its coming had brought an end to hauling goods and people across the desert from Shoshone, there was also a great deal of resentment about the heavy-handed, monopolistic way in which it was run. Not only did people dislike the slowness of the trains, they also grumbled about high rates, running behind schedule, accidents and breakdowns.

Laying track on the Minidoka and Southwestern. Many of the railroad workmen were immigrants from Japan which, at the time, was at war with Russia. When Admiral Togo won a great naval victory in 1905, it was said to have spurred the immigrants to lay steel "at a record breaking pace."

Soon after construction of the Twin Falls irrigation project began, the Land and Water Company had sounded out the Harriman interests about running a branch line across the tract. In response, engineers of the Oregon Short Line—the division of the Union Pacific that ran through Idaho—surveyed a possible route for such a branch. After a great deal of fraternizing with the railroad officials by Burt Perrine, Bob McCollum, Walter Filer and others, the railroaders agreed to the proposal.

In January, 1904 the "Minidoka and Southwestern Railroad Company" was incorporated to build and operate the branch. Starting at Minidoka, the line would run southwesterly and cross the Snake River several miles northeast of today's Burley. From there, it would continue downstream to Milner, thence across the tract, generally conforming to the course of the river.

Earlier, officials had said the company would begin laying track in the spring. But, just after the Minidoka and Southwestern was formed, this plan was overruled, apparently by Edward Harriman, himself. The president of the branch line, E. E. Calvin, announced that construction would not begin until it was certain that settlers were taking up land on the Twin Falls tract.

This foot-dragging by Harriman resulted in the loss of many months and, for a time, an angry Frank Buhl considered canceling his request for the line. But cooler heads prevailed and Buhl's subordinates convinced him to get the project started by agreeing to make good the railroaders' costs should the irrigation project fail. Accordingly, a contract was made with the Minidoka and Southwestern, which required that the dam and first 25 miles of canal be completed before June 1, 1905. If not, the company would be given a one-year extension, but would have to pay the railroad 6% interest on the money it had spent for construction. If not completed by June 1, 1906, the Land and Water Company would pay the railroad the interest plus $100,000 in damages.

Work was finally begun on the branch, by the Utah Construction Company, in November, 1904 and, eight months later, on July 28, 1905, the tracks reached Twin Falls, and locomotive Number 619 with Engineer Carter at the throttle, entered town.

To celebrate the occasion, the town held a "Railroad Day" on August 8. A large crowd listened to speechifying by Idaho's Senator Weldon Heyburn, and Representative Burton French. Another orator, Fred Reed, praised the "pioneer women of Twin Falls for their courage in leaving comfortable homes and making another in the sage brush." Reed also led three hearty cheers for President Theodore Roosevelt, "the friend of reclamation and the friend of the West."

Various field sports and contests were held on Main Street East—foot races, quarter mile horse races, a pie eating competition, and others. Harry West of Kimberly won a $4 pair of shoes for coming

Daily passenger train at Twin Falls depot, ca. 1906.

in first in a fat man's foot race, and a baby contest was won by Blanche Harrah, the first child born in Twin Falls.

The people were served a free barbeque in the partially-completed Hotel Perrine. Donated beeves and sheep were strung on rails and roasted in huge pits, across Main Street from the hotel. A grocer, H. F. Harder, baked 400 loaves of bread, and Burt Perrine furnished two wagon loads of Blue Lakes fruit for the feast. Barrels of well-iced lemonade were available on every corner of Main Street. After the feast, the Pocatello city band gave the town its first ever band concert in front of the hotel and, in the evening, free dances were held at two locations.

The established train schedule allowed Twin Falls passengers to take one of two daily runs to Minidoka, where they could connect with east- and westbound trains on the main line. One was a regular passenger train, and the other a freight train that had one passenger car.

This latter run was known variously as the "accommodation train," the "ping-pong," or the "grasshopper." Both it and the passenger train stopped at all the stations, but stops were longer for the grasshopper because freight had to be handled. Since the "grasshopper's" passenger car had no smoking compartment, men who wanted to light up had to visit a box car at the end of the train which was said to oscillate like "a jig saw in motion."

That autumn of 1905, freight traffic was very heavy. Forty carloads or more were arriving each

The Depot

Immediately after the railroad was completed to Twin Falls, work was begun on passenger depots for most towns along the route. The Twin Falls depot was situated on the corner of Sixth Avenue South and Shoshone Street South, where today there is a parking lot for the Depot Grill. Agent Dan Sullivan took occupancy in February 1906.

A freight shed, and a stockyard were also erected in the rail yard, and a turntable was installed to turn locomotives back around in the direction of Minidoka.

Another view of depot, ca.1906.

day, and a large number of them contained the personal effects and furnishings of the many farmers who had spent the tract's first crop season working their land, and who hadn't had time to deal with their goods. Now they were able to build homes, convert their prove-up shanties into stables, and engage in farming and commerce, unrestrained by inadequate transportation.

Ira Burton Perrine with display of local agricultural produce on "Railroad Day," 1905.

Twin Falls "Electrified" By Completion of Shoshone Falls Power Plant

The Great Shoshone & Twin Falls Power Company's electric plant at Shoshone Falls came on line for the first time on Saturday, August 17, 1907, brightening peoples' lives and making their day-to-day tasks much easier.

Shortly thereafter, the Presbyterian Ladies Aid Society of Twin Falls sponsored a "Merchant's Indoor Carnival," at Billy Ring's roller skating rink. As part of the entertainment, some of the ladies costumed themselves and performed "representation drills" in which they portrayed some of the town's businesses.

Shoshone Falls power plant and penstock, ca. 1907

Mrs. Bedford, the druggist's wife, made a great hit depicting the power company in a routine called "The Electric Girl." Over her dark costume she had looped strings of lights, powered by copper conductors fastened to the floor. The effect was regarded as "very pretty indeed."

"The Lady Barber" was also well received by the audience, as was "Painless Dentistry," in which a woman represented the use of Dr. Lewers' new electric drilling machine.

The Twin Falls *News* reported that the 850 h. p., 500 k. w. General Electric turbo-generator at Shoshone Falls was "purring like a satisfied tomcat." It was driven by water fed through a 420-foot long, 10-foot diameter tunnel from above the falls, which provided a pressure of 80 pounds per square inch. Steel gates at the tunnel's upper end controlled the flow, and water was carried from its base to the power house through a steel penstock supported in cement cradles. The plant had been designed so that additional generating units could easily be added to produce more power when required.

Readers may recall that the project had been conceived six years earlier by Harry Hollister and Ira Perrine, the latter of whom held the water right. A 2.6-acre parcel, on the north side, was needed to accommodate the power house, a tunnel and several diversion dams between the islands above the falls. This land was thought to lie within state-owned Section 36, and, in December 1901, the land board gave Perrine and Hollister a lease on the property, with an option to buy it after making certain

Interior of power plant, ca. 1907 showing the first generator.

improvements.

But, just to the east was Section 31, containing the Shoshone Falls hotel, which was owned by a group of wealthy investors. Due to differences between survey lines on the north and south sides of the river, these investors were able to claim that part of the land Perrine and Hollister needed for their project belonged to them.

William A. Clark, leader of the hotel property owners, argued that the power plant would destroy the natural beauty of the river and falls and would thus degrade the tourist business.

Fearful of the effect the dispute might have on his lease, Hollister, in early 1902, filed lawsuits in district court at Shoshone, against the state and Clark interests, seeking to condemn the required land, as a public right of way, by proceedings in eminent domain. In turn, the Clark interests filed for an injunction to restrain Hollister from performing any further work

The Idaho press and general populace were overwhelmingly in favor of the power project, so it came as no surprise when juries approved the condemnation, and granted the hotel owners and the state only token sums in damages. The result was further bitter litigation—suit and counter suit, and appeal after appeal—that continued for years.

Clark, a Democrat from Montana, had been elected in 1899 as a U. S. senator. But, just after taking office, in 1900, the Senate had found him guilty of bribery and corrupt election practices, and declared his election null and void. Clark vacated his seat but was once again elected to serve in the Senate from 1901 to 1907.

As the litigation continued, Idaho newspapers routinely berated Clark, calling him a "hide-bound miser who prostituted politics in his own state." If he was worried about beauty at the falls, it was said, he should "spend two bits" and paint his "ramshackle...red barn" of a hotel," or lease it to someone for a long enough period to warrant making improvements.

The wrangling frightened away one group after another of Hollister's power plant investors, most of them Chicagoans or New Yorkers, known to Hollister through his various mining ventures. The Shoshone Falls Power Company was incorporated by one group, then re-incorporated by another.

Undaunted, Burt Perrine fed the press a constant stream

Not the First

The electric power plant at Shoshone Falls was not the first in southern Idaho.

A smelter near Ketchum installed a small system in 1882, and, in 1887 Hailey became the first town in Idaho to have a power plant.

Boise got a small unit, in July 1887, that used water from the Ridenbaugh Canal.

J. H. Brady's 1,500 h. p. plant at American Falls began providing power to Pocatello in July 1902.

Even Ira Perrine had a small generator at the Blue Lakes, beginning in 1902, that he used to run fruit dryers.

Twin Falls *News* advertisement, 1907.

of puffery—one big financier or another was about to join in the venture; power would be available in just a few weeks; an electric rail line would run between Twin Falls City and the Falls; other plants would be built at Auger Falls, the Twin Falls, and in the Hagerman Valley, and so forth. Perrine was stimulated to push the project even harder when the Twin Falls city fathers gave him the concession to provide electricity to the town for 50 years.

Finally, in November 1906, substantial financing was found, and just a few months later, the brothers J. S. and William S. Kuhn acquired a major interest in all of the Snake River power projects. Reorganized as the "Great Shoshone & Twin Falls Power Company" the Kuhns soon had the Shoshone Falls system completed.

For several years a small fossil-fueled generating unit had been lighting the Hotel Perrine, and a few nearby businesses, but it was routinely shut down at midnight. On that momentous night when Shoshone Falls electricity first went on line, the town remained brightly illuminated until dawn. So enthralled were citizens that a large group clustered under the new arc lamps and refused to go to bed until 4 A. M., when the chief of police called their attention to the time.

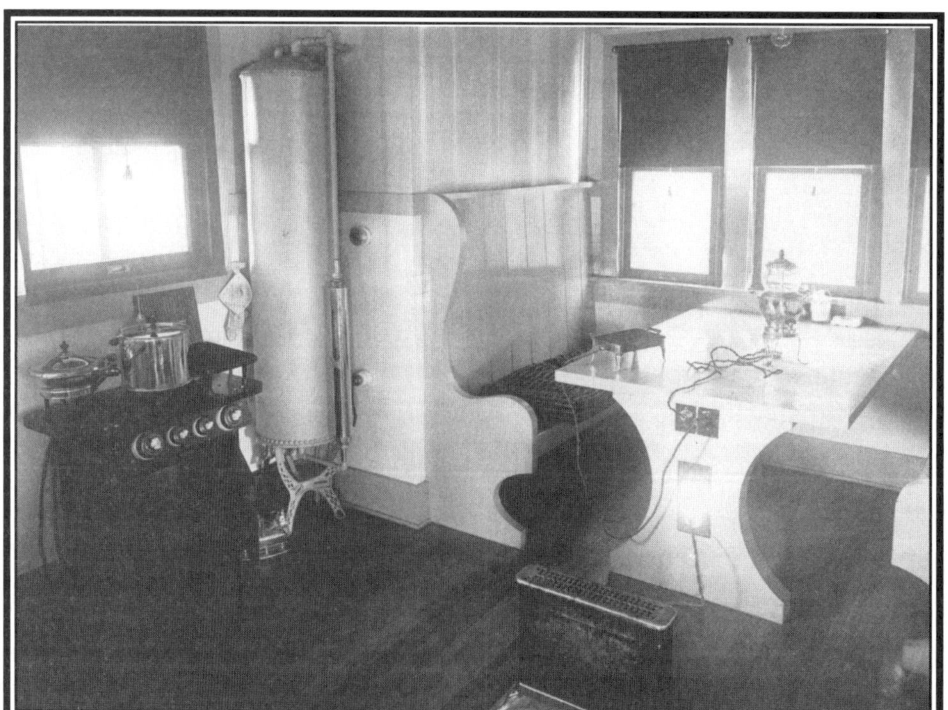

Early Twin Falls electric kitchen. Power cost 17 cents per kilowatt-hour.

"Smoke Wagons" Come To Magic Valley

—Beginning of the End for Horse and Buggy Era—

Charley Cline and "Mexican Joe"—the cowboys tending Point Ranch, 30 miles south of Twin Falls—had just finished their supper on a dark and stormy May, 1907 evening. Charley was doing the dishes, and Joe was strumming a sentimental tune on his guitar, when a strange sound suddenly filled the air. "Sounds like a stampede," said Joe, as he headed for the door, "but there's no stock around here." "Beats me," said Charley as he hung up the dish cloth.

Then, a flash of light came through the ranch house window, which Joe thought might be a thunderstorm approaching. When the two wranglers went to the door and opened it they were blinded by an intense, unrecognizable beam of light. According to report, Joe began muttering prayers in Spanish while crying that it was a "volcano spewing lava," and "he and Charley were goners." Just then Joe's horse came to the door, whinnying pathetically, and, before Charley could bar the way, the animal rushed inside the house.

The strange light and throbbing sound grew nearer and nearer until, at last a dark green object was seen to sweep around the corner of the corral and stop in front of the house. The volcano, it seems, was simply the first automobile ever to come to that vicinity. The operators of this "stink wagon" were three men from Butte, Montana who were on their way to the mining town of Contact.

While Charley got some supper for the travelers, Joe threw a pitchfork full of hay in front of the green monster, and then returned to his guitar.

A. M. Sande, owner of Twin Falls Feed and Ice, in his auto, ca. 1910.

"Autocars" were very much a novelty in the United States for the first few years after Twin Falls' founding. A handful were in use in Pocatello, Hailey and other established towns, but it wasn't until May of 1905 that Magic Valley saw its first, when entrepreneur Fred Reed brought one to Burley for use on an "automobile express" route between that town and Oakley.

Reed claimed that the mules of Burley intended going on strike at the first whiff of gasoline, and had agreed not to haul any disabled "smoke wagons" that might fall by the wayside.

> **Dan Patch**
>
> In1904, the year that Twin Falls was born, the horse was king. Just as a fine car is coveted today, so was a fine Percheron stallion or draft animal in those times.
>
> Heaven was watching the team-pulling contest at the county fair, or the horse races on July 4; Hell was being told your horses had glanders and had to be destroyed.
>
> Dan Patch was the most popular animal in America. That year the famous pacer ran the fastest mile ever made by a horse in harness, completing the race in one minute 56 seconds flat. Two years later, as a nine-year old, he cut the time to 1:55 ¼, a record that would last for 32 years. Dan Patch never lost a race.

On January 11, 1906 a lawyer named Marlin Sweeley brought the first auto to Twin Falls. When the one-cylinder "runabout" was being unloaded, a frightened horse ran away, and was said to have crashed through the window of a barbershop, much to the astonishment of its occupants.

There were only several hundred miles of paved roads in the entire United States, so cross-country trips were high adventure, and certainly not for the faint of heart. The prudent motorist always carried materials for repair of flat tires, as well as a winch, a rope and a deadman to haul himself out of trouble.

Today we can jump in our cars and whiz to and from Boise in several hours each way. But when James Clinton and Craig Coffin drove here from the capital city in late October, 1906, en route to Salt Lake City, it took them four and a half days. Their route, much of which was over very uneven roads, deep with sand, took them via Orchard, Mountain Home, Clover Creek, Bliss, and Hagerman. From the latter place they took the north side of the river to the Blue Lakes, where they crossed on Perrine's ferry.

The car was reported to have "shied a little on the ferry and showed a disposition to duck" the spray at Perrine Coulee, but otherwise behaved splendidly, climbing the steep south side grade at four miles an hour. The run from the top of the grade to the Hotel Perrine in the center of town was negotiated in seven minutes. The travelers went on to Salt Lake City and, upon their safe return to Boise, the editor of the Twin Falls *News* declared that their journey had established the practicability of autos as road machines.

Setting out on an automobile trip to Swiss Valley, near King Hill in 1912. This picture was probably taken at the Blue Lakes. The man on the running board is believed to be I. B. Perrine and the apples adorning the car are probably from his orchards.

But there were many people who refused to admit that any machine could ever replace the horse. Said one national farm magazine: "The automobile has been making considerable noise the past few years and cutting up capers and people, while the horse has been quietly...eating oats...Horses have been steadily going up, which is better than 'blowing up' as the automobile has been doing...Automobiles may come and go...but his majesty the horse remains monarch of the road...No machine ever built or still un-invented will successfully dispute his reign...not while the automobile remains blind to your actions of kindness and dumb to the sound of your voice, nor while the horse is the delightful company he is..."

More and more people in Twin Falls bought cars over the next two years. One of them was the mayor, Fred Voigt who in July, 1908 ordered a 50-horsepower Northern road car and then began showing an unusual interest in the condition of the roads on the tract. It was, perhaps, no coincidence that his associates on the city council passed an ordinance, about the same time, limiting automobile speeds to eight miles per hour within the city limits. The editor of the Twin Falls *News* nonetheless advised citizens to keep their dogs and chickens out of the road until Voigt "gained steerageway."

1905 Buick Model B.

Early in 1909 the city got its first auto dealership when Carl Lind started selling Buick "Tourabouts" at $1050, F.O. B Flint, Michigan.

Studebaker buggy ad in 1906 Twin Falls *News*, when autos first came on the scene.

A Day at the Circus

In Twin Fall's infancy, a full-fledged circus coming to town was a grand and exciting treat. Our town saw its first such entertainment—Cozad's "big trained animal circus"—in September 1905, two months after the railroad reached here.

The second to play Twin Falls was the well-known Sells-Floto Circus, out of Denver. It was early in the 1907 traveling season—Friday, May 24—when the troupe arrived and its outfit cars were switched onto a spur track adjacent to the designated circus grounds.

One can be sure that many enlivened and anxious youngsters were on hand at daybreak next morning to watch the animals being cared for, the huge canvas tents being spread and raised, and their stakes being driven home with sledge hammers. Some of larger boys might even have hired on as temporary laborers.

Ad in T. Falls *News* May 17, 1907.

Sells-Floto would be in town only one day, giving two performances, but that day would provide more excitement than anyone could have imagined or wished. The afternoon was a warm one, and the hardy Twin Falls pioneers basked in the sun after a hard winter in their rude tents and shack-like homes. Before entering the "big canvas" for the first performance, they strolled past a free "grand street display."

Some of the 100 "Imperial Amazing Acts " to be seen inside the big tent were "daring" acrobats, "champion" bareback riders, a "Ben Hur herd of Arabian stallions," 24 clowns, and a "Royal Japanese" troupe. And, just for the men, there was a "congress of beautiful women in physical culture," performing "exploits of perfect equilibrium" and "artistic posing."

Also touted was the huge menagerie. According to report, the "old world" had been "scoured thoroughly for wild animals and strange beasts." Just how wild would soon become evident.

After the afternoon performance, a large crowd gathered inside the menagerie tent to watch the animals being fed. The Bengal tigers were restless and at the sight of the meat, they threw themselves against the bars of their cages and roared.

Markel, the largest and most beautiful tiger in the menagerie, had always been under control and was not considered vicious. But his keeper failed to pull down the sliding door on Markel's cage, leaving an open space at the bottom through which the animal thrust his paws, and, in an instant, had pushed up the door and sprung out onto the back of a Shetland pony.

The tiger sank his teeth into the jugular vein of the little

Sells-Floto Circus Parade in 1908

horse as the keeper seized an iron bar and struck the feline's head. Markel released his hold on the first pony and seized another, from which he was driven by a second blow with the bar.

Meanwhile, there was a stampede inside the tent. Panicked men, women and children fled in every direction. Some climbed on top of the cages, others crouched behind wagons and many more escaped under the canvas. One man found himself directly in the path of the beast as it turned from the ponies and started for the entrance of the tent. The man crouched and looked the tiger in the eye and it turned aside and passed him, leaping straight at Mrs. S. E. Rozell and her four-year old daughter Ruth, who were both borne to the ground and mauled by the beast.

W. H. Bell, with his wife and children, had been standing alongside Mrs. Rozell. Without hesitation, Bell placed himself between his family and the tiger, drew a .32 caliber Smith & Wesson pistol from his pocket and opened fire on the enraged animal.

The tiger had sunk his fangs into the little girl's neck while Bell was drawing his weapon, but the first bullet caused it to open its mouth. Bell moved closer to the big cat and fired again. The animal turned his head toward the shooter, growling and lashing his tail, but Bell coolly placed a third bullet behind the animal's shoulders.

Releasing his victims, the tiger started west alongside the Pioneer Storage Company's warehouse. Bell followed, firing three more times. At the sixth shot the tiger turned and started crawling back toward the tent, but collapsed and stretched out, stone dead.

The entire incident had taken only a few minutes. Mrs. Rozell arose, her wounded daughter clasped to her breast. The two were hurriedly taken to a doctor's office, surrounded by a large crowd of concerned citizens. While the mother hadn't been seriously injured, Ruth had a broken neck and was bleeding profusely from a torn jugular vein. Within two hours, she was dead.

Bell was lionized (tigerized?) for his heroic confrontation of the maddened animal. Afterward, he was the only person on the grounds who could give a coherent account of what had happened. He said he normally didn't carry the pistol, but some unknown impulse had prompted him to slip it into his pocket before leaving home. A modest man, Bell said he had only done what was necessary to protect his family. He had felt no fear because there hadn't been time to think about it.

Little Ruth's funeral was conducted several days later at the Methodist church, the Rev. W. H. Parker presiding. There wasn't much concern about grief counseling in those days—six tough little girls, dressed in white, bore the white casket into the crowded church. Afterward, the funeral procession rode to the Twin Falls cemetery, as the bell tolled at Bickel School.

Warrants were issued for the arrest of the circus owners and manager, and the keeper of the tiger, on charges of manslaughter. But, apparently, the entire affair was settled a month later when the circus managers gave the Rozzells $1,750 for "expenses" in connection with Ruth's death. They also later sent one of Markel's claws to Bell, in recognition of his daring act. Certain citizens of Buhl, where Bell lived, thought he was entitled to a Carnegie medal for his act, and they were going to make an effort to secure it for him.

Sells-Floto Circus parade of May 27, 1911.
Note steam calliope.

The First County Fair

—No "Grafters" Were Allowed—

In announcing that the county's first agricultural fair would take place from September 5th through 7th, 1906 the Twin Falls *News* told its patrons to:

> "Paste these dates in your hat. If you wear cuffs, write the dates on [them]. Tie strings around your fingers, tell your wife to remind you. If you haven't a wife, tell your friend's wife to jog your memory...No admission fee will be charged and no grafters will be allowed within gunshot of the grounds."

The concept for this wholesome annual tradition had its inception the previous year, near the end of the first growing season on the newly-irrigated Twin Falls tract. At "Railroad Day" —a celebration marking the completion of the Minidoka and Southwestern's tracks to Twin Falls—prizes were awarded for a variety of agricultural exhibits.

Produce display at first fair, in 1906.

Also, a few farmers had taken samples of their husbandry to the state fair at Boise. Having experimented a great deal that first year, they proudly displayed a wide variety of produce, including such unlikely items as peanuts, sugar cane, cotton, broom corn, and tobacco. The overall exhibit won $100 for second place in the sweepstakes. Everyone came home, all fired up, with plans to hold a Twin Falls fair the following year.

Alexander McPherson, the Twin Falls Land and Water Company's agricultural expert, got things going by soliciting nearly a thousand dollars of startup money from his employers, and others. The Farmer's Federation took charge of the planning, and by the summer of 1906 had appointed superintendents for sixteen competitive divisions, and had published a list of premiums (prizes) and instructions for preparing exhibits.

A fairground was chosen right smack in the middle of town, in the block on which now can be found The Ballroom, U. S. Bank, and other businesses. This site was picked because there were two small frame buildings there that had been used for the previous year's school classes. For $425, the Twin Falls Land & Water Company bought these buildings from the school trustees and loaned them for use in displaying exhibits.

The sagebrush and weeds were cleared from the grounds and, using donated lumber, volunteers built a bandstand, stalls and pens for livestock, and coops for poultry. Then, the entire area was decorated handsomely with bunting and hundreds of flags.

Ad for Jewel Range in Twin Falls *News*, September 7, 1906.

Using a free water connection provided by the waterworks company, the ladies of the Women's Christian Temperance Union fitted up a women's restroom in a brick building next to the *News* office, which was then at 153-155 Main Avenue West. No mention can be found of any arrangements made for men.

In addition to the premiums, cash prizes were announced by the Land and Water Company for the best- irrigated field of ten or more acres, and for the best acre of a variety of different crops

Special prizes were also offered by several businessmen. The Studebaker carriage dealer promised a "runabout" worth $100 for the best individual display of agricultural products.

The most interesting proposal was that of Clark Gibson of the Twin Falls Implement Company, who said he would present a "Jewel" range, to the first couple to announce their willingness to be married in public at the fair grounds, and who lived up to their promise. Gibson guaranteed that the "Jewel" would "burn sagebrush and...not burn bread," and claimed that as an "agent of domestic harmony, it was unexcelled."

There were "many estimable young bachelors" — said the male-chauvinist-pig editor of the *Twin Falls News* – "who were wasting half their time in trying to cook for themselves and keep their houses free from dust," and there were "many charming young ladies who might be willing to assume domestic responsibilities." If any such young folks showed up to get married and claim the prize, however, they went unmentioned.

The fair opened as scheduled and was a great success. For the first time, the hard-working pioneers of the tract could show the fruit of their industry, and the fecundity the newly-tilled, virgin soil.

Fair-goers could see livestock, produce and handicrafts brought in from locations as distant as Roseworth and Murtaugh. There were displays of handmade bricks, harness, tinware, willow chairs, carvings, and needlework. There were huge sugar beets, potatoes, and other vegetables.

Interesting as the displays were, the amusements fell somewhat flat. The daily concerts of the newly-formed Twin Falls band were well received, but some of the other events didn't materialize. A planned cowboy relay race was canceled because all the good riders were busy with the fall cattle roundup. A scheduled jiu jitsu match between Japanese wrestlers fell through, and some of the baseball clubs that were scheduled to play in a tournament didn't show up

A baby show was arranged at the eleventh hour, it having been found impossible prior to that time to find any judges brave enough to officiate. Within a few minutes of announcing that there would be a contest, 31 entries were made. Later, Twin Falls' only jeweler gave cups to the winners.

A variety of saddle and trotting horse races were held, as well as some boys' foot races. But the most popular outdoor feature of the fair was the horse team pulling match. Sixteen farmers proudly showcased the animals upon which they relied for their livelihoods. Every entry pulled the same

Trotting race at first Twin Falls county fair.

load— a farm wagon with its wheels chained, on which were seated a number of men. The way the entrants "gee-d, haw-d, perspired and kicked up dust" was said to be "a caution." The event was won by Joe Nielsen, a farmer and grading contractor who had worked on the canals, whose team of white horses pulled seventeen men.

The fair continued to be held in Twin Falls until 1916, when it was moved to a permanent location at Filer, purchased with $13,000 in solicited funds. The event has been held each year since 1906, except for the depression years of 1932 through 1934, and the war years of 1942 through 1944.

Judging horses at 1910 county fair.

Early Twin Falls Journalism

Forerunners Of *The Times-News* Were Unashamedly Partisan Papers

The salutation in the initial issue of our town's first paper—the *Twin Falls News*—bragged that, within a few years, the new little village of Twin Falls would "rank with the foremost of the smaller cities in the west." It sat, after all, smack in the middle of "largest body of irrigable land in the United States," where the climate and soil were ideal for growing many crops.

Co-owners Charles P. Diehl and O. H. Barber began publishing the weekly paper on Friday, October 28, 1904. A subscription only cost $2.00 per year —if paid in advance—and that price included a periodic magazine called the *Irrigation Age*.

TWIN FALLS NEWS

VOLUME 1. TWIN FALLS, IDAHO, FRIDAY, OCTOBER 28, 1904. NUMBER 1

TWIN FALLS CANAL PROJECT

Brief Outline of the Progress and Purposes of the Most Extensive Irrigation Enterprise in the United States, Watering Finest Tract of Land in the West—Dam Across Snake River an Engineering Marvel and Canals and [illegible]

The greatest irrigation project in the United States, and the third largest in the world, involves an expenditure of approximately $2,500,000 and the reclamation of 271,000 acres of the best agricultural lands in the great Snake River Valley of Southern Idaho. The tract was formerly set aside for a national park, because of its wonderful scenery. By a deal with the state officials, who are taking advantage of the Carey act of 1894, making a donation of 1,000,000 acres of arid lands to each state that will construct canals and have the lands reclaimed, the great park will become a garden of small farms and vineyards.

The enterprise includes the construction of two irrigation canals and laterals which will have a combined length of over 1,000 miles. The main canal is 26 miles long, 80 feet wide at the bottom and 80 miles of canal 40 feet at the bot- [illegible]

to secure the rural mail system. Public telephones and all modern conveniences are to be added to the comforts of those making homes under the canals. Settlers have now located land to the extent of 40,000 acres and are clearing, fencing, plowing and building homes awaiting the coming of the water in the ditches. The company has built the main canal and the dam is nearly completed, 60 miles of laterals covering 30,000 acres are completed and the company announces that the work will be pushed as fast as possible until every acre of the large tract has been placed under the canal. Water rights are to be sold to actual users at reasonable rates. This amount of lands will support a number of good towns when once under cultivation.

RIVERS BURIED UNDER LAVA.

The Snake River, or Shoshone River, has its head around the [illegible] collection of springs, called the Thousand Springs, of 5,000 second feet flow, falls 200 feet over the lava walls. The Snake River has also had its troubles, and while it spread out and went over the lava above American Falls and down as far as the head of the great canal, it finally got the best of it here, and at the three islands where the dam or rather three dams are about completed the water broke through the lava and boiled and sputtered and fell into the large crack of lava while cooling, and now it falls in jumps of 10 to 200 feet at a jump for 60 or 80 miles, leaving a mesa from 100 to 800 feet above the bottom of the Snake River canyon. This mesa, or what is known as the Snake River Valley, is a vast plain of sage brush, without a break of any great depth, with lava ash soil and sandy loam, making a most fertile foundation for any kind of fruit or cereal, which grows in great abundance when the soil is softened by moisture.

THE DIVERTING DAMS.

To get water up to the level desired to irrigate the land, it is necessary even here to dam the river and raise its normal level about 49 feet [illegible]

GROWTH OF TWIN FALLS CITY

First Improvement was a Seventy-Mile Telephone Line from Shoshone to Twin Falls and Milner—Water Works Constructed—Various Lines of Business Represented and Many Improvements Inaugurated—Railroad [illegible]

On the first of April there was not a sagebrush removed from the site of Twin Falls City. The first improvement that came was a telephone line, south from Shoshone to Blue Lakes, Shoshone Falls and Twin Falls City, thence east to Milner, a distance of about seventy miles. The first building erected was the office of the Twin Falls Investment Company, which was first occupied July 27. Perrine & Burton's store, which was the second building erected, was the first occupied, having been opened for business July 23. On August 10 Prothero's restaurant was opened, to be followed at short intervals by other business institutions. At the present time there are two general stores, two livery stables, two restaurants, two saloons, a lumber yard, brick yard, meat market, blacksmith shop, rooming house, and a real [illegible]

135 voters, and many who have not established a residence.

Such, in brief, is Twin Falls City today. Enough business establishments already here or under way, to supply present demands, but plenty of room in the near future. It is not the people now here, but those who have come, located land and gone away, and will return with the coming of water, augmented by the continuous stream of people coming for the first time which will make this the most populous and prosperous section of the state within an incredibly short time.

The mighty Snake river under absolute control; with canals completed covering a large tract of land; a wonderfully productive soil suited to a wide range of products and especially adapted to fruit culture; a delightful climate; settlement by a thrifty and educated class of people; im-

Front page of first issue of Twin Falls *News*.

Despite the initial puffery, editor Barber took pains to ensure his readers that the *News* was in no way connected with the Twin Falls Land and Water Company, or any other interest or industry. It was in no sense going to be "a professional boomer." Nevertheless, every edition of the paper described the wonders of the tract and pointed out business opportunities.

***News* printing plant, in early 1905.** On the left is Chas. Diehl.

After only about four months, on February 25, 1905, Barber and Diehl dissolved their partnership and Diehl became the sole editor and publisher. Barber went to his old home town of Salt Lake City, bought a press, and came back to "founder"—as Diehl said—a second weekly paper, the Twin Falls *Times*, on March 23, 1905.

Diehl only lasted a few weeks on his own at the *News*. By the April 7 issue, he had sold out to the "News Printing Company" and had moved to the new town of Burley, where he soon started up another paper.

The News Printing Company, consisted of William M. Wayman from Payette, George F. Sprague—the treasurer of the Twin Falls Investment Company—and George B. Fraser, who took over editing the *News* on April 1, 1905.

Fraser was a former city editor of the *Boise Statesman*, and had also been a publicist for the Boise company employed by the T. Falls Investment Company to sell land on the irrigated tract.

Readers of this column may remember Fraser as the intrepid newsman who fell sound asleep while listening to a lengthy speech by the "Great Commoner," William Jennings Bryan, in 1907.

The *News'* new owners were, from the start, unabashed "boomers." At every opportunity they promoted and advertised the tract and new little community of Twin Falls.

In the summer of 1905 they built a new 25 by 60 foot building, with a basement, which is believed to have been located at 155 Main Ave. West. The latest model linotype

Twin Falls News Not The First in The Area

In that hot summer of 1904, when our town of Twin Falls began abuilding, there were already a number of newspapers in southern Idaho. In those days even small towns might have more than one paper. At one time, Hailey had four—two dailies and two weeklies.

The two papers nearest to Twin Falls were the *Journal* and the *Herald*, at Shoshone, some 25 miles to the north, and the *Albion Times*, even farther away to the east.

Robert McCollum had owned and edited the Shoshone *Journal* for several years before teaming up with Burt Perrine, in the latter's grand enterprises.

Interior of second *T. Falls News* office, occupied about Sept. 1, 1905. George Fraser stands at the left with his favorite hunting dog. Like most early papers, the *News* did commercial printing. At its office it sold stationery and forms, shown here in the display cases.

machine was purchased and installed in the new building.

Linotypes, which combined the tasks of casting lines of type, and setting them, had been around since 1886, but the *News'* new Merganthaler machine, with its two magazines, enabled a compositor to change back and forth instantly from one face or font to another, even in the same line.

Under George Fraser's control the *News* prospered; by 1907 the paper could boast a circulation of over 2,600. The rotund editor spoke forcefully and clearly on a variety of community issues—bonds for a sewer system, fire protection, an adequate waterworks, to name but a few.

Like most editors in those days Fraser had not graduated from any journalism school. Many of his kind were practical printers, who had served as apprentices in that trade. Others were simply graduates of the school of hard knocks. William J. Hill, an earlier editor of the *Owyhee Avalanche,* at Silver City, had been a miner, a compositor, then a keeper of the Owyhee Ferry, before trying his hand at the editing game. It was said his body was scarred all over from the piercings of Indian arrows and musket balls.

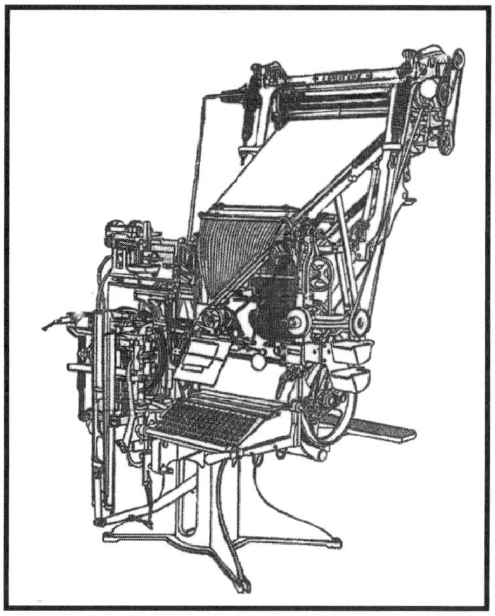

The News' improved linotype.
(Author's collection)

Fraser and most of his contemporary editors were likewise tough as nails, and possessed a wry,

First *T.F. News* office, on Main Ave. S. in May, 1905. Posing with his old stage coach, is driver Charles C. Haynes, a veteran of the Overland Stage Co. and many other early lines.

self-denigrating humor. No colored ribbons ever adorned their offices, nor dangled from the sagebrush out in front. If they were alive today, most would be widely shunned, and labeled as being bigoted and chauvinistic. But, in their day, they were immensely popular with their contemporaries.

Unlike those today, that claim to present a balanced view of the news, in that earlier era, newspapers were blatantly and unashamedly partisan. The *News*, *Times* and the *Shoshone Journal*, for example, were initially Republican sheets, and ardent supporters of President Theodore Roosevelt. The *Idaho World* at Idaho City and the Silver City *Avalanche* were two of those of the Democrat persuasion.

The eight-page *News*, like its competitor, the *Times*, was a lively and uninhibited little journal. Both papers contained a minimum of national and international stories, since neither would obtain a wire news service until World War I. The emphasis was on local events—parties and outings, house guests, trips to town and such mundane happenings.

There were also a lot of filler items, such as tips for housewives, romantic short stories and serials, poems, travel articles about foreign lands, and anecdotes with a moral. Also, in the earliest days, every edition contained a piece dealing with animal husbandry, irrigation and other farming subjects.

Advertisements promoted a host of potions such as Dr. Williams's Pink Pills For Pale [anemic] People; Lydia Pinkham's Vegetable Compound for menstrual pain and ovarian problems; Hall's Canker and Diptheria Remedy, and children's Castoria—a cure for constipation, sour stomach, diarrhea, worms, convulsions, and loss of sleep.

O. H. Barber sold the *Times* in 1906 and Fraser left the *News* in 1910. Over the ensuing years, a number of individuals owned and edited the two papers, and they appeared variously as dailies,

T. F. *Times* office, ca. 1908.
Wilbur Hill, who owned the paper from 1907 to 1916, is at far right.

twice-weeklies and weeklies.

In 1932, Roland B. Tofflemire formed a partnership with Emil Bordewick, the then owner of the *Times*. Together, they bought the *News* from Roy Read, who had owned it since 1913. Starting on January 1, 1937, both papers were published daily under the same roof—the *Times* in the evening and the *News* in the morning. Then, on February 16, 1942 Tofflemire and Bordewick consolidated them as the *Times-News*.

Twin Falls Public Library

—A Rich History of Citizen Involvement—

The Twin Falls library, with its collections exceeding 173,000 items, is one of the finest in the West. And, like many others of the period, it came into existence as the result of volunteer initiatives. Although , early on, the Twin Falls Land and Water Company had set aside the southeast corner of the Twin Falls city park for a library, nothing was done until March, 1906, when a committee of the Owl Club—an historical and literary organization—began agitating for a library, and decided to "open negotiations" with the philanthropist, Andrew Carnegie, who had already established hundreds throughout the country.

In June, the club produced the play "Charlie's Aunt," and earmarked the proceeds to buy books. But, with little else having been accomplished by April, 1907 the Twin Falls *News* began serving as the local outlet for the state Traveling Library, which periodically brought cases of books from Boise and left them for a time to be checked out, at no charge.

Mrs. Joseph H. Seaver, who lived south of town, used her pony cart to pick up book donations and contributions for rummage sales.

The paper provided this service until February 26, 1909 when—thanks to the efforts of Mrs. Warren H. Barnes and certain other Twin Falls women—15 citizens formed a library association.

Tradition has it that Mrs. Barnes began cooking a few extra pots of her special Boston baked beans each week, which were sold for her at the bakery run by the brothers, Zenas and John Smith. When she had earned $10, Mrs. Barnes called a meeting of interested parties in Kennedy Packard's real estate office where ways of starting a library were discussed, and the association was formed.

Using the bean money and contributions from other charter members, two rooms were rented on the second floor of the Harder building at 117-119 Main Avenue East, each illuminated only by several overhead electric light bulbs. Businesses and individuals donated lumber for shelves, chairs, heaters, electric power, and books to get the project started. This first rude home town repository was opened on March 7, 1909, loaning books, for a one-time fee of 25 cents, to locals who could "furnish a guarantee of honesty."

A series of rummage sales were held over the next few months, and large contributions were received which allowed the association to order $200 worth of new books. By September, six hundred volumes were on hand, and a certain Miss Goetz was serving as librarian.

When the Library Association met in early December it rescinded the 25 cent fee and made the use of the library free for all townspeople. In order to raise funds it was decided to start up a "Penny-a-Day Fund" to which every local household would be asked to give one cent each day. The Syringa club—later called the 20th Century Club—was to collect these pennies monthly.

Furthermore, as part of what it considered its educational responsibility, the Library Association decided to sponsor a series of weekly lessons to study Esperanto, a so-called "international language" that had been created in 1887 by a noted Russian linguist. The lecturer would be Floyd B. Hardin,

Children celebrating Book Week in 1926 in front of library in the McCollum Building on Shoshone St.

who had been hired in November to replace Miss Goetz as librarian, and who claimed to hold the "highest professor's certificate from the Esperanto Council at London."

Hardin's first lecture on the trendy subject was given January 15, 1910 in the Knights of Pythias Hall. He is reported to have used sterioptican slides to illustrate his presentation, and to have had on hand a "remarkable collection" of Esperanto literature and letters from Esperantists in 40 countries.

Although an interest in Esperanto still exists, our down-to-earth Twin Falls pioneers weren't a bit taken with the complex language. After several lectures, the librarian-lecturer resigned his position, said good-bye and left town for elsewhere. Or, as we say in Esperanto, "la bibliotekisto-lectoro abdiki la lia ofico." Mrs. Sadie Rogers, his replacement as librarian, only spoke English.

About this time, supporters began pushing for the city to step up and provide a regular appropriation to run the library. Editorials pointed out that most cities supported theirs with taxes, while Twin Falls had let a few individuals do the work which properly belonged to the municipality.

But no governmental action was taken, so the Library Association, along with service clubs and volunteers kept things going by sponsoring frequent benefits. A "poverty dance" was held at the Dreamland Pavilion and, at a "Dance of the Books" — called the "party of the decade" — attendees came representing book characters.

The most spectacular of these benefits was a grand musical and dramatic production called "Songs of the Nations" held in 1910 at the Methodist church. The performers included a full-blown orchestra, a "Welsh chorus," the Methodist choir, and leading vocal solists of the city. Part of the program included an account of the Battle of Waterloo, and an oration on the *Marseillaise,* said to be "the greatest war song ever written."

Finally, in 1912, the library was designated as a city department, with a yearly budget of $1500. This triggered renewed efforts to obtain a building grant from the Carnegie Foundation. But a misunderstanding between the parties involved resulted in the request being rejected.

Over the years the library was moved four times: to the Light and Power Building on Main Avenue (1910); the new county courthouse (1912); the McCollum building at 227 Shoshone Street (1913); and the Elks building on Shoshone Street, now known as The Ballroom (1927).

Not until the coming of the Great Depression would the library get a permanent home. With the promise of a supplemental grant from the federal government, in 1938 voters approved a bond issue to build on a site that the city had purchased earlier with money donated by a local businessman, Louis I. Benoit. The following year, the library opened at its present location.

Community Support Continues.

In 1986, after certain state and federal funding programs were eliminated, causing the library to lost part of its book budget, a group of citizens incorporated the non-profit Twin Falls Public Library Foundation.

The purpose of the foundation is to raise money from private sources for library support.

In recent years the foundation has contributed about $63,000annually to the library, most of it going to provide about 25-35 percent of the budget for printed and audio books.

The library, built in 1939, as it appeared before completion of a large addition.

Kimberly's Earliest Days Enlivened By Social Activities

On May 15, 1905, the pioneers of Kimberly had a party. The occasion was the arrival of the first building in the new, un-surveyed little town. Water had just come onto the virgin land, and irrigating had begun in earnest, but, nonetheless, over 50 farm couples from the surrounding area took part in a good, old-fashioned "hoe down" at the Stockgrowers' Mercantile Company's store.

Never mind that the building — which had been hauled from Milner — had been set up in the center of what would become a street, and later would have to "continue its journey a little farther" to get it on the proper lot.

At the party, food cooked over a sagebrush fire was served, after which the Twin Falls Land and Water Company's agricultural "superintendent," Alex McPherson, spoke to the farmer's wives about

Kimberly Main Street, looking south, ca. 1909. *(Courtesy Kimberly Public Library)*

dairying and gardening.

Dancing began at 10 p. m. to melodies played by the Swydenski orchestra from Twin Falls. Kimberly "Mayor" James C. McMillan had supposedly agreed to dance the Highland fling with his irrigating boots on but, if he did so, it went unreported.

A few weeks earlier, McMillan had settled six miles east of Twin Falls, and declared that a new town site, called Kimberly, would be platted on his 160 acres. McMillan, a former resident of Illinois, had just resigned his job as private secretary to Idaho's new governor, Frank R. Gooding. Earlier, he had worked for the Land and Water Company, promoting the Twin Falls tract.

McMillan's new townsite was located on the line of the railroad which at the time was not far to the east, building rapidly toward Twin Falls.

In mid-July a townsite company was incorporated with McMillan its secretary and general manager. Other directors of the corporation included Thomas and Fred Gooding, both brothers of the governor. The Gooding boys were successful Shoshone area businessmen and ranchers, and both of

The Farmers' Association

For a time, the Stockgrowers' Store continued to be used as Kimberly's principal meeting place, especially by the Farmers' Association, which organized June 7, 1905, with William S. Starr as its president. A wide range of farming subjects was taught at its meetings—"Raising Sheep on the Ranch," "Measuring Water," and "Poultry on the Farm," to name just a few.

A school teacher, Miss Brittomart Wolfe, was a Kimberly land owner and an active participant, as was James McMillan's daughter, Jean.

The association provided many opportunities for socializing — box dinners, dances, hay rides, etc. — but it was also active in dealing with irrigation and road problems. On one occasion, Jim McMillan was chosen to take a road petition to the county seat at Albion, and he agreed, provided someone would loan him an airship (blimp) to get him over the unbridged ditches.

them had taken up land adjoining McMillan's Kimberly townsite.

By July 22, when the railroad reached it, the new town was being surveyed, its streets were being cleared and the irrigation laterals were being bridged. The town had a restaurant and the Scott brothers' Kimberly Drug Store was about to open. A stagecoach was making regular runs and, within a week, scheduled passenger trains began running both east and west.

A formal opening sale of townsite lots was held on August 8, with a good crowd present, many of them brought by special trains from Minidoka and Twin Falls. For the refreshment of the assembled buyers, the townsite company provided sandwiches and lemonade, and the Kimberly ladies sold ice cream and cake, raising over $25 to help buy an organ for the school to be opened in the fall.

Depot with local train alongside, ca. 1906. One observer called it "handsome" and the "most conspicuous building in Kimberly." In 1973 it was cut into three sections moved and rebuilt, and now serves as the town's library. *(Courtesy Kimberly Public Library)*

Peter Kimberly in Twin Falls, 1905.
(Courtesy T. Falls Public Library)

A "Beautiful Character."

Peter Lanterman Kimberly died suddenly in Chicago of heart disease on June 4, 1905, just a few weeks after the town of Kimberly was named in his honor.

Three years earlier, he had joined with his good friend, Frank H. Buhl, to provide most of the money for building the Twin Falls irrigation project. At the time, both of them were awash in cash, having recently sold their steel mills. Kimberly was a former resident of Buhl's home town, Sharon, Pennsylvania.

Kimberly, who had visited our area several times, was considered by many locals to be a "beautiful character," whose chief aim seemed to be "to make people happy."He knew what it was like to be poor, having twice met "the wolf" and prevailed.

"I have more money than I need," said Kimberly a few weeks before his death, and, if he could put it where it would do some good, he would feel that the Almighty had caused him "to prosper for a purpose."

The drawing for order of choice in lot selection was held at the drug store, where a churn was provided as a receptacle for name cards. The cards were thoroughly mixed and then drawn out, one at a time, by five-year old Oliver West. Asking prices ranged from $100 for residential lots to $750 for the choice business locations. More than fifty sales were made that day.

Perhaps Kimberly's most lively party in its first few years was the reception held to mark the opening of a railroad depot, on August 16, 1906. Among the 300 or so people attending was a large group from Twin Falls, including that town's brass band.

Upon arrival, the Twin Falls people took off their coats and made themselves right at home. Jim McMillan, upon seeing the *Twin Falls News* editor, George Fraser, and several other portly gents, ordered the cooks to 'Make at least 800 more sandwiches." Soon the band had unlimbered and "the rafters of the new depot reverberated the echoes of good music."

Fraser reported that, because of the depot's space limitations, he and all the other corpulent gents were relegated to doing their dancing outside on the railroad siding.

The frolic continued until midnight when refreshments were offered at the Mercantile Company's store. The Twin Falls people were served first so that they wouldn't miss the train taking them home.

At one a. m. the conductor called "All Aboard" and, while the guests complied, all of them cheered for Kimberly and everything connected with the town.

The combination passenger and freight coach selected for the excursion proved to be too small to hold the returning crowd, so the conductor coupled on a coal car in which the men rode. Editor Fraser would have his readers believe that these fellows enjoyed a comfortable ride, since the car was clean, and the wind was just right to carry the locomotive's cinders away. Sure.

Kimberly's First School. Opened in October, 1905, the building was erected by popular subscription on a lot donated by the townsite company. It was replaced a year later by a new brick structure. The teacher, Miss Nettie Durphy, at far right, owned 40 acres and was an active member of the Farmer's Association. *(Photo courtesy Gene Gillette, grandson of Kimberly pioneer, W. H. Turner)*

"Buster Brown" Visits Twin Falls To Peddle Shoes

Mary Jane, Buster Brown and Tige in 1905. Brown Shoe Co. gave the name "Mary Janes" to the single-strap shoes that both Buster and Mary Jane wore. The name soon became a general term for strap shoes, a staple in the wardrobe of most little girls. *(Author's Collection)*

Celebrity promotions are not unique to our modern era. Back in 1907, when Twin Falls was a mere three years old, a local department store brought to town one of the most popular commercial mascots of the day. And, when "Buster Brown" arrived. it was said that, "in proportion to his size," he attracted more attention in than would have President Theodore Roosevelt. "Without a doubt," claimed the promoters, "this was the greatest advertising attraction" ever brought to the city.

"Buster Brown" was a charmingly-mischievous young cartoon character created by Richard F. Outcault, an illustrator who had previously published other strips called "The Yellow Kid"and "Poor Li'l Mose."

From the day he first appeared in the *New York Herald*, May 4, 1902, Buster, together with his dog "Tige" and sister Mary Jane, delighted children of all ages. Indeed, Buster's Lord Fauntleroy clothes and antic behavior became as familiar to those who read the "funnies" at the turn of the last century, as were Little Orphan Annie and Charlie Brown to succeeding generations.

The somewhat sissified Buster was always carefully dressed, with his long blond hair in bangs. His constant companion, "Tige" (Tiger) was said to be the first talking pet in American comics, though the cartoonist never let adult characters in the strip observe the bulldog's speech.

Buster's pranks usually resulted in a little homily meant to set the young reader straight on right and wrong.

By 1904, "Buster Brown" had become so popular that John A. Bush, a sales

William Ray as "Buster Brown," on the road in 1905.
(Courtesy Brown Shoe Co.)

Idaho Department Store, ca. 1907. This sponsor of the Buster Brown promotion, was a descendent of one of the first general stores to open in Twin Falls, G. W. Tarr and Co. In July, 1904 , Tarr occupied a tent near Rock Creek, then moved to a frame building on Main Ave. In March, 1905, Tarr was bought out by Harry F. Allen of Boise, and others, doing business as Allen Mercantile Co. In June, 1906, Allen sold out to his partners and the business was renamed the Idaho Department Store. Occupied in September 1906, the building shown above was at 158-154 Main Ave. S. Offices and a hall on the 2nd floor were used by the Knights of Pythias and by the first officials of the new Twin Falls County.

executive with the Brown Shoe Company, came up with the idea using the character as a symbol for the company's line of children's shoes. That year, Bush bought the right to use the "Buster Brown" name from Outcault, for $200.00,

Working swiftly, the Brown company then introduced Buster Brown Shoes to the public at the World's Fair — the same "Meet Me In St. Louis" fair that introduced hot dogs. Buster was played by a 41-inch high, 44-year old actor, with a squeaky voice, named "Major" William Ray, who had convinced the company that he could better portray the comic character than a real child. Ray's bulldog became Tige, Buster Brown's dog. Ray had previously been a circus performer.

So popular was Ray's routine at the fair that, for the next 26 years, the shoe company used

Ad in Twin Falls News, May 7, 1907 .
(Author's Collection)

Buster Brown "reception" in 1905 at Jonesboro, Ark.
(Courtesy Brown Shoe Company)

him and other midget actors to play Buster Brown in promotions all across the United States. Well-promoted in advance, they toured the entire country pitching Buster Brown shoes with performances in theaters, department stores and shoe stores.

It was Saturday, May 11, 1907 when the William Ray and his traveling show reached Twin Falls to advertise Buster Brown Blue Ribbon shoes for children on behalf of the Idaho Department Store. He and Tige had just visited all the large Eastern cities and were on their way for a tour of the Pacific Coast.

Despite Ray's train being somewhat delayed, he was met at the Twin Falls depot by a large delegation of children and their parents, whom he greeted with childish effusion, despite his apparent years. The Idaho Department Store had a "tallyho," drawn by three large horses abreast, to carry him. While Ray was taken to the Hotel Perrine for a meal, three local boys, each with a megaphone, scurried around town to announce the time and place of the "reception" (performance).

Ray was scheduled to make his pitch in front of the department store, on a platform that had been built just for the occasion. But, since it had rained all night and morning, making the streets "fully sloppy," the event was held at Billy Ring's Skating Rink..

At the reception Buster—who was described by the *Twin Falls News* as an "exceedingly droll...quaint little chap...." — cracked jokes while he explained in detail how the shoes were made, and displayed the different grades of leather used in them. Tige did his part by performing stunts, such as untying the shoelaces of any child not wearing Buster Brown shoes. The dog was a replacement for the original Tige who was said to be "laid up with a bad case of rheumatism." While all this was going on, some 2,000 souvenir buttons were passed out which showed Buster and Tige "in one of their favorite attitudes."

In a letter written several days later from Boise, "Buster" complimented Twin Falls children for being an orderly group, and he bragged that, at the finish of his pitch, he had had "every child cheering and promising to buy the Buster Brown shoes."

Buster Lives!

John A. Bush, who came up with the idea of using Buster Brown as a mascot, went on to become president of Brown Shoe Co. in 1915, and Chairman of the Board in 1948.

Bush continued to promote the brand using print, radio and outdoor advertising. Buster was also the subject of a popular series of silent film comedies from 1925-29. Later, talkies were also made.

In 1926 the Buster Brown Radio Club was on the radio twice a week. From the mid-1940s to 1952 a network carried "Smilin' Ed's Buster Brown Gang,"which subsequently moved to television.

The Brown Shoe Company still exists and continues to market Buster Brown shoes for children, as well as other lines.

Twin Falls' Premier Photographer

—Life and Art Were One for Clarence Bisbee—

Clarence Bisbee, with his photo equipment, not long after he arrived in Twin Falls.

Back home in Nebraska, his family called him "Babe." His wife dubbed him "Bee." Today, we know him as Clarence E. Bisbee, the photographer who seemingly took pictures of every dam, ditch, farmhouse, crop, and stunning landscape in our region.

As a young man, Bisbee engaged in farming until finally deciding to make his favorite hobby—photography—his profession. Several years after graduating from an Illinois college specializing in that subject, a friend he had met there urged Bisbee to leave the Midwest and come make his fortune in the booming new town of Twin Falls. That friend was Charles Diehl, who had recently founded the *Twin Falls News*.

Bisbee was 30 years old when he arrived here, with his equipment, on January 27, 1906. A partner, J. Reichardt, arrived soon afterward with his outfit, and, by the end of February, the two of them had established a studio (and a home) in a 16 by 28 foot tent on what is now Second Ave. East, across from today's Blacker's Appliance and Furniture. At first, sagebrush fires provided heat, and water was carried from the nearest hydrant or ditch.

Bisbee bought out Reichardt's interest in the business in mid-August, 1906 and continued living and working in his tent. But, a year later, the power plant at Shoshone Falls came on line and, by mid-November, 1907, he had opened an "electric studio" on the ground floor of the Commercial and Savings Bank at 112 Second St. East, across from today's Rogerson Mall.

With electric lights, said the *Twin Falls News*, he could now "set the sun at defiance" by taking pictures "at all hours and under all climatic

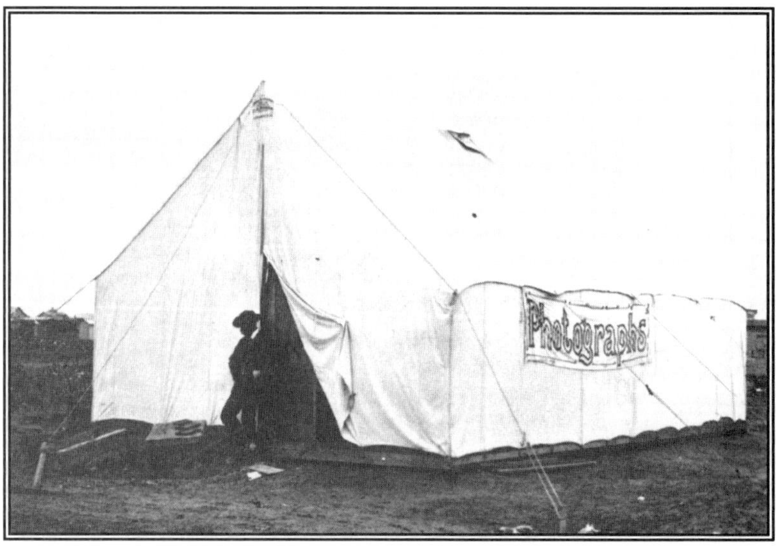

Bisbee's tent studio and home in 1906.

conditions." It didn't matter whether the day was cloudy or clear, the picture was taken just the same. Bisbee pressed the button and "the Shoshone Falls power plant did the rest."

Bisbee's business flourished in the new studio and in a branch of the business that was open in Buhl for a time. He did portraits and weddings, sold color pictures of Shoshone Falls, and began providing photos for the newspapers. But most of his income came from providing the illustrations for brochures promoting the various agricultural tracts. Using valley scenes, he also made and sold postcards.

Bisbee had become engaged to a Nebraska school teacher named Jessie E. Robinson before he left for Idaho, so his social life in Twin Falls was mostly restricted to male companionship. He was active in the Y. M. C. A. and in the Masonic and Modern Woodmen brotherhoods. He was also treasurer of a prohibitionist group and ran,

Bisbee' s Photographic Tools

Throughout his career, Clarence Bisbee always used a tripod. His camera of choice had an 8 by 10 inch lens, but he shifted to 5 by 7 when doing postcard pictures.

Most of his exposures were made using the dry plate negative process, in which a gelatin was used as the means of dispersing light-sensitive silver salts on a glass plate. About 1878 the dry process had superseded an earlier wet plate method, which had been used as far back as Matthew Brady's time, and which required immediate processing after exposure, before the plate could dry out.

The dry plates, such as Bisbee used, could be bought, ready for use. The photographer did not have to treat the glass, merely expose it to light and develop it.

Bisbee and many other photographers continued using glass plates well after photographic film came into wide use.

Bisbee's 1914 studio. Note the aphorism "Life and Art Are One" on the face of the building.

The Bisbees pose with glass negatives and post card stock.

unsuccessfully, in 1909, for the city council on the temperance ticket.

In the spring of 1910, Jessie quit her teaching job, came west and married Bisbee in a Baptist church in Salt Lake City on June 15. After the newlyweds returned to Twin Falls, the 26-year old pitched right in and helped her husband with the business. She would later recall often walking with him ten miles a day "carrying an 8 by 10 outfit and sometimes a 5 by 7 too, with all extra plates and camera cloths."

The Bisbees built and opened a new, modern brick studio in October, 1914 on the southeast corner of Second Ave. East and Second St. East (now Hansen St. E.). A small one-bedroom apartment in the rear of the building became their new home.

As the masons were laying the sidewalk in front, Bisbee imbedded in the cement an iron case containing a picture of his original tent studio, which had stood just 75 feet south of the new building.

The S. R. Sullivan twins—Veryl and Verdon—were the first to be photographed in new building and this happy occasion was celebrated with a grape juice toast to the future of the babies. Bisbee and his wife would never have children of their own, and they always displayed a fondness toward those of others.

Bisbee continued to prosper, though less of his work was in making promotional pictures. But, the coming of the Great Depression brought bad times to many families, including the Bisbees. In the early 1930s their business dropped off, and their investments in real estate ventures caused a huge financial drain.

One of their properties was a parcel along the Shoshone Falls grade, which had several small lakes, in which they tried, unsuccessfully to grow fish for the commercial market. They also searched the property, in vain, for valuable minerals. Camping on the place, however, afforded them great pleasure, including the *al fresco* dinners eaten there four Christmases in a row.

In 1936, at age 52, Jessie died at home in their

The Bisbees at one of the fireplaces in their new studio. Though not visible in this picture, the fireplace is inscribed: "God's in his heaven, all's right with the world." The other two displayed the aphorisms: "Trifles Make Perfection" and "Perfection is No Trifle."

apartment. A despondent Bisbee, his business failing, retired in 1939 and, sometime afterward, his studio was sold, perhaps as a result of a mortgage foreclosure. He moved onto the Shoshone Falls property and lived in poverty there and other places until his death on June 3, 1954. Both he and Jessie are buried in the Twin Falls cemetery.

After the photographer's passing, three local men—O. A. (Gus) Kelker (a former editor of the *Times-News*), DeWitt R. Young and Dr. Wallace Bond—each chipped in $500 to buy about 2,300 glass negatives from Bisbee's estate, and later donated them to the Twin Falls Historical Society. Today, our public library has custody of these wonderful plates, which comprise such an important chronicle of our early history.

Not The First To Arrive

Clarence Bisbee wasn't the first photographer to come to Twin Falls. One of the state's best – Earnest Tacha – came from Mountain Home and bought some land north of town. In March, 1905 he and his wife, Hetta, began proving up, and he opened a studio on Main Street, a full 10 months before Bisbee arrived in town.

Soon, Tacha's "eloquent" pictures began appearing in the *Twin Falls News*. Among these is the well-known January 1906 picture of the "Homeless 20" group.

Another photographer already in Twin Falls when Bisbee arrived was I.. L. Young. Many farmers hired him to photograph their fields and homes, and his pictures also frequently appeared in the *News*.

Leah Amos. One of Bisbee's later competitors, she opened a studio in Oct., 1907 above the Post Office, two doors west of the Hotel Perrine.

Some Notable Twin Falls "Firsts"

Initiation goat cart as pictured in sales catalog. The wheels of this model produced a clatter like a galloping horse. Moving the handle up and down made it buck. The manufacturer also suggested using one of their throwing blankets in conjunction with the goat ride, so that candidates might "experience the thrills of a 'Double header.'" *(Courtesy Phoenixmasonry Masonic Museum – www.phoenixmasonry.org)*

As this column approaches its conclusion, it seems appropriate to list several "firsts of a kind" in the history of our town and valley, that haven't been mentioned elsewhere.

Riding The Goat.

These days, fraternal organizations are going the way of the buggy whip but, in early Twin Falls, just about everyone belonged to one or another of the lodges that soon sprang into being. They were a way to make friends, form bonds, and, perhaps, attain a bit of refinement through meaningful rituals.

The first group to organize was the Independent Order of Odd Fellows, which was instituted March 27, 1905, two days before the Woodmen of the World formed their lodge.

Not all of the Odd Fellow rituals were dignified or meaningful. At one of its early celebrations at the skating rink, the daughter of a member got roars of laughter from the crowd for reciting, "When Papa rode the goat."

Her poem referred to a stunt that was part of the lodge's initiation ceremony. First, the initiate would be taken into an anteroom where he was blind-folded and doused with denatured alcohol, so that in the event of a fatal accident, he would be immediately recognized by St. Peter—himself an Odd Fellow—and be admitted into Heaven.

After being anointed, the candidate was hoisted onto a wooden goat cart, and was then wheeled wildly around the lodge room and dismounted in front of the "Noble Grand" for a nonsensical interview. After this, one presumes, the more serious part of the initiation took place.

Start of Odd Fellows 1906 anniversary celebration parade at Twin Falls' new depot. Shoshone Silver Cornet band in the lead.

Keeping The Peace.

Scott W. Hazen, proprietor of the Twin Falls Rooming House, was the first constable for the village, beginning in November, 1904. Two months later, the Cassia County Sheriff appointed Frank Ramsay (who ran the city water company) to be the first deputy sheriff for the Twin Falls precinct.

But, crimes were rare in the earliest months, the peace officers being mostly concerned with stopping the "reckless discharging of firearms" in town, and in rounding up "egg-sucking" drifters who raided hen houses, or robbed traveling men while they slept. Later on, there were gambling houses, opium dens and "red-curtained" houses of prostitution to raid, and the occasional horse thief, forger or murderer to track down.

Tradition has it that, in the spring of 1905, a cave was driven a few feet into the east wall of Rock Creek Canyon for use as a jail. The few men who were supposed to have been put there were told to keep an eye peeled for rattlesnakes.

The first real jail building was completed in April 1905 and declared ready for the first individual who "made himself obnoxious." It was built where Krengel's hardware was formerly located, and had three cells crammed into a space of 16 by 19 feet. Its first tenants were " John Doe and Richard Roe," who were charged with being drunk, discharging a gun and failing to pay for a meal.

First Religious Services.

According to C. S. Walgamott, the first sermon in what is now T. Falls county was delivered on February14, 1883 by a Methodist minister named Jocelyn, who was spending a few days at the Shoshone Falls Hotel. Using a slab of lava about 5 feet high for a platform, the Rev. Jocelyn delivered his message to a small audience.

On July 14, 1904—about the time the first buildings were going up in town—Rev. Henry W. Parker, of the Shoshone Methodist church, became the first minister to visit the town of Twin Falls. That afternoon, he went to Perrine's ranch for dinner and to secure a free lot from the Townsite

Company, upon which to build a church. Seven other denominations would also ultimately receive such lots.

Parker was also the first minister to hold a "preaching" church service in Twin Falls. It took place on October 1, 1904 at the Hazen Hotel, with 41 persons present. After the little wooden school house was erected in late November, it was used by most denominations for their services, until they could erect their own buildings.

Twin Falls Catholic Church, completed late October, 1905.

Christian Church in about 1906.

First Churches.

The Catholic Church was first to complete a regular house of worship in Twin Falls, as reported in the *Twin Falls News* of November 3, and December 1, 1905. It was located at the corner of 2nd Avenue East and Fifth Street East.

The Christian Church was second to be finished. An evangelist, Rev. LeRoy F. Stephens—who, like his Lord, was an excellent carpenter, and who, it was said, had built more churches than any man in the country—arrived in Twin Falls and, within a week had raised enough money to build a small chapel of his design. It was completed about November 17, 1905. The church owned a lot donated by the Townsite Company, but the building was erected on two purchased lots near the park.

Memorial Day.

In the year 1904 there were still quite a few veterans alive from a deadly conflict fought decades earlier. These survivors of the Civil War were mostly all members of the Grand Army of the Republic, which held annual "encampments" nationally and in the various states.

Many of these old men, along with younger Spanish-American War veterans, participated in Twin Falls' first observance of Memorial Day—or Decoration Day, as it was then commonly called—held on the afternoon of May 30, 1905, in the little temporary school house.

The vets occupied a place of honor in the front seats, and proudly wore their organization insignias. Two of the old timers present could boast of having shaken the hand of martyred President Abraham Lincoln.

Surrounded by flags and sprays of wild flowers, the Rev R. D. Osterhout offered a short prayer, and then spoke to the audience about "Our Honored Dead." Charles Cole, a representative of the "Sons of Veterans" spoke about how the loss of his father in the Civil War had affected him and his family. He reminded the audience of "the duty of every true American to aid in maintaining a nation where freedom and liberty may be enjoyed, though the cost of such a land was dear." To conclude the ceremony, a male quartet sang "Tenting Tonight.

Fred Reed's Town

A"Child of the Desert" Opens Burley Townsite

Fred Reed sitting behind Elva McCollum in "Homeless 20" picture, 1906.

Burley is "Fred Reed's town," said a newspaper editor in February 1906. Reed was once described as a person "in the habit of doing things"—a "newspaper man, cowboy, promoter, traveler, empire builder, entertainer, prince of good fellows. In short, a character, a man of ideas and of ability. A man who believes in Idaho...who is devoting his...energy to her upbuilding."

It's not surprising that "Major" Fred R. Reed was the one chosen to get Burley up and running.

Two factors—the creation of the Minidoka Irrigation Project, and the coming of the railroad—led to the building of a town in the lower Goose Creek valley. On November 1902, the federal government created the Minidoka tract, in accordance with the terms of the recently-enacted Reclamation Act. Surveys were begun on both sides of Snake River, and in April, 1904, the secretary of the interior allotted $2.6 million to build a dam, canals, an electric power system and pumping plants to get water to land not susceptible to gravity irrigation.

Families began settling on project land that summer and, in October, railroad bed grading was begun for an 85-mile long branch of the Oregon Short Line, that would connect the towns of the Twin Falls irrigated tract with the main line at Minidoka. This road—the Minidoka and Southwestern—would run from Minidoka to Snake River, where a bridge would be built, thence across the Twin Falls tract, ending at Buhl.

The first mention of what would become Burley appeared in the *Shoshone Journal* of November 18, 1904, when it was announced that a railroad siding would be built at the "town site on the river." In early January, 1905 it was confirmed that such a town was to be built on approximately 640 south side acres, encompassing the old Jordan ranch.

By early April, John E. Hayes was surveying the new town site, which lay about a mile southwest of the railway river crossing. Hayes was the same fellow who had laid out Twin Falls at a 45-degree angle.

Two weeks later, articles of incorporation of the Burley Townsite Company were filed with the secretary of the state at Boise. The incorporators were I. B. Perrine of Blue Lakes, Josiah E. Miller of Albion, S. D. Burton of Milner, William T. Jack of Oakley and S. H. Hays of Boise. Major Reed was named general agent of the company.

The opening sale of lots was scheduled for May 1, so Reed— known to his friends as the "child

Townsite Ad, 1905.

of the desert."—got right to work, furnishing, as one observer noted, "the hot air and ink" to give the locale "notoriety." He dubbed Burley the "Ideal City," pointing out that it would surely become the natural depot of supplies for the Minidoka Tract, Oakley and Albion, as well as the center of river navigation between Minidoka and Milner dams.

Burley—so Reed's advices claimed—had a pure and abundant water supply from a well driven to a depth of about 45 feet; a long-distance phone line was completed and a telegraph line was being strung from Minidoka. A hack line ran daily to and from Twin Falls and soon there would be a stage connection to Minidoka, Albion and Oakley

Reed bragged that several business houses had already been constructed just outside the townsite limits that would move into town as soon as lots could be acquired; that an attractive railroad depot would soon be built; also that, before long, a brickyard would be in operation. His only complaint dealt with the "coyote band," comprised of some drunken tent dwellers with great strength of voice, who had gotten into the habit of giving nightly concerts.

Although the railroad bridge hadn't been completed,

Opening Day Drawing for Burley lots, May 1, 1905.
(Courtesy Idaho State Historical Society)

the opening sale came off as scheduled on Monday, May 1, 1905. Two special trains arrived on the north side of the river that morning carrying prospective buyers from various points in Idaho and Utah, and bringing dining and tourist cars to provide eating and sleeping accommodations. The steam tug "Mary McLean" and the gasoline launches of Walter Filer and S. D. Burton ferried people across the river and took them on excursions.

Burley Depot, ca. 1907
(Courtesy Cassia County Historical Museum)

The first edition of the *Burley Bulletin,* printed by Charles P. Diehl on April 28, was available for perusal by the visitors but, unfortunately, it was only a half-page in size because of Diehl's trouble in getting paper delivered by what he called the "Jim Crow" stage line run by "shell-pated cigarette fiends."

With 1,500 people present, Reed and his assistant, George Redway, began the sale at 1:30 p. m. About 500 people drew for lots and made purchases, totaling about $50,000.

The sale was deemed a great success and, afterward, the townsite company directors entertained a group of guests with an oyster supper, in a restaurant which had just been built that very day.

A few weeks later, the railroad bridge was completed, lending impetus to Burley's growth. The depot and an opera house were completed that year, and plans were made for a wagon bridge to be thrown across the Snake at the foot of Overland Ave., to connect Burley with the north side portion of the Minidoka tract.

But, despite the growth, it was tough going in those earliest days. Within its first 18 months, the town suffered a disastrous fire that destroyed eight buildings, and a windstorm that

Burley's Namesake

The city of Burley was named in honor of David E. Burley, one of the incorporators of the Minidoka and Southwestern Railroad.

Once described as being full of "push, vim and energy," Burley was hired by the Union Pacific Railroad in 1879, and when the O. S. L. was created as a subsidiary to the U. P., he was made its general passenger agent. He was exceedingly popular, in part because of his generosity in doling out free railroad passes. He had more than once saved the editor of the Silver City *Avalanche* from "either counting the ties or staying at home."

Burley got his first glimpse of his namesake town in May, 1905, shortly after its opening day. That year he also acquired a quarter-section of land just east of Buhl, and had an architect build him a fine house upon it. Later, he moved into this "prove-up shack" to fulfill his Carey Act obligations.

When he bought a brand-new incubator, mounted on wheels, his Buhl neighbors suggested that, henceforth, the busy executive could set the eggs, take the device with him on his longer business trips, and expect to have chicks upon reaching his destination.

D. E. Burley home near Buhl, about 1907.

demolished houses and barns, and moved a "substantial" two-story building off its foundation.

And, the surface of the land, it seemed, was always either deep in dust or sticky mud. In one early baseball game between the local "Ditch Diggers" and the Twin Falls "Irrigators," it was said the earth was so wet that a Civil War veteran who had lost both his legs at Shiloh was afraid to attend the game, lest his wooden stumps take root and sprout. When Burley, the "city of destiny," lost the game 6-13, Major Reed was said to have wept such a puddle behind first base that he had to be removed from the field in S. D. Burton's gasoline launch.

Fred Reed soon went on to other tasks — the building of American Falls and the Twin Falls northside project, to name just two — but Burley continued to prosper without him, hampered only by the failure of the government to get Minidoka project water on the land when promised.

David E. Burley
(Courtesy Cassia County Historical Museum)

Magic Valley's First Airplane Flight

-- Strange Potato Discovered After Forced Landing ---

The people of our area saw their first "aeroplane" flight just about ten years after Orville and Wilbur Wright made their historic heavier-than-air flight at Kitty Hawk, North Carolina, and about nine years after Twin Falls had sprung into existence.

The flight took place in the fall of 1913, as part of the "Great Oh-ad-i Festival"—also known as the Southern Idaho Fair and Carnival. The main feature of the three-day "Oh-ad-i" was the regular county agricultural fair, but it also included other wholesome diversions, such as city band concerts, baseball games, dances at the corner of Main and Shoshone, and a "grand automobile parade."

But, without question, the flight of Dr. Frank J. Bell's fascinating flying machine proved to be the most spectacular event of the fair.

Frank J. Bell in his Curtis Biplane.
(From T.F. News photo Sept. 13, 1913).

Bell—who was described as a noted airplane expert—had come down with his Curtis biplane from his home in Billings, Montana to perform. It isn't clear how he got the machine to Twin Falls but, to avoid long cross-country flights, it probably came, partially disassembled, on a railroad flat car.

Bell was scheduled to fly on Friday and Saturday, September 18 and 19, taking off and landing from the Twin Falls baseball field.

The field, complete with grandstand, was just then home to the Twin Falls "Magicians," an amateur baseball team that played in an independent southern Idaho league. Bell was to take off immediately after the conclusion of a Magician's game against the semi-pro Ogden Utah team.

All of those on the grounds, who had paid 50 cents to see the ball game, were allowed to remain to see the flight free of charge.

That Friday afternoon of September 18, the weather was perfect for a flight, and a huge crowd was present at the ball field. "Bird man" Bell started promptly on schedule, taking off in his "splendid machine" from immediately in front of the grandstand.

For about thirty minutes, so said the Twin Falls *News,* he made "spirals, circles and glides enough to convince the most skeptical of the value of the flying machine." The plane sometimes reached a thousand feet in the air, where "the slight figure of the doctor, balanced on his small support, seemed only part of a great bird, which was pursued in its steady flight by real birds attracted by its unusual appearance. Aside from the steady hum of the big eight cylinder motor, which drove the plane, there was little difference in the flight of bird and man-driven machine."

First airplane to fly at Twin Falls, September 1913.

But, when Bell attempted to land his plane in the small ball park, his speed was too great to permit a safe touchdown. Passing only 50 feet above the heads of the spectators, the plane sailed on westward and disappeared amid the house tops.

Bell's intent was to return for another try at landing in the park, but a slight leak in the engine's water tank forced him to land the plane unexpectedly in Ben Ott's potato field, just west of town.

Baseball field, located at 2nd Ave. S., between Ketchum S. and Liberty S. (now the Latham Motors property).

Bell knew all about landing amongst potatoes. Once before, at a frontier day celebration, he had tried to land at too fast a clip on a small ball field, like that at Twin Falls, and had banged spectacularly through a high board fence, plowed through a cluster of Indian tepees, before finally coming to a stop in a spud patch.

The aviator left his plane in Ben Ott's field overnight and, on Saturday afternoon, he tried to get it in the air for his next performance. However, after several attempts he was obliged to use a nearby road for take-off. What with all the telephone and electric light poles skirting the road, it was a hazardous job, but Bell succeeded in getting up, and gave his huge audience a perfect flight, ending with a smooth descent into the ball park.

A few days later, farmer Ott found a potato in his field, near the spot where the airplane had landed, that bore a "startling resemblance" to the plane. The tuber was 8 inches long and, extending from each side — like on the plane — were two long branches.

The potato was put on display at the Cress grocery store. Said the Twin Falls *Times*, tongue in cheek: "The only moral to this story is that farmers should be more careful in allowing strange objects in their potato fields when tubers are in the ground, as there is no telling what may happen."

From Potato Field to Modern Airport

On September 4, 1927— fourteen years after Frank Bell landed in a Twin Falls potato field—Charles A. Lindbergh flew his "Spirit of St. Louis" low over Twin Falls' city park and dropped a scroll bearing greetings to the townspeople. Several thousand had gathered to see the famous "Lone Eagle" who had, in May, flown solo from New York to Paris.

Lindbergh was making a three-month tour of the nation, and was en route from Salt Lake to Boise, where he made his only landing in Idaho. Twin Falls, at the time, had only a crude 100-acre airfield.

The following year, Lionel Dean, an aviation enthusiast, bought himself an open-cockpit biplane. To provide a safer place to land this first aircraft to be owned by a local, Dean made a 1,320 foot dirt runway on his property south of Kimberly road, and east of the East Five Points.

By 1929, with community interest on the rise, 240 acres, south of town, were obtained for an airport. Two runways were made there, one of them about twice as long as Dean's.

But, this airport ground sloped badly to the east so, after World War II, the city obtained an 1800-acre site to the northwest of it. Here a new, all-weather regional airport, called Joslin Field, was opened in 1948.

The field's name honors S/Sgt. Raymond R. Joslin, a local airman who was killed in September 1942, while serving with a bomber squadron in the Solomon Islands.

Trolley Folly

—Burt Perrine's Electric Railway—

When Ira Burton Perrine's electric street cars arrived in Twin Falls, on April 29, 1913, he declared that one of his "fondest hopes...had been realized," and that he was sure his planned trolley line would be a resounding success. Events, however, would prove him woefully wrong.

Only a few years after arriving in Idaho, young Burt caught a bad case of the 'trolley fever" that was sweeping the country just then, as cities everywhere began running cars powered by overhead electric wires, or slotted conduits in the roadway. In fact, Perrine was very likely part of the "Shoshone Falls Power and Electric Railway Company" that was formed, in 1890, to harness the immense water power of the Shoshone Falls, and use it to light Shoshone Junction, and run an electric railway for tourists from there to the Falls.

Nothing came of this grand plan, but, from 1901 through 1911, when the Shoshone Falls power plant and the various irrigation projects were being built, Perrine's plans for an electric railway appeared in his every pronouncement. At various times, it was claimed that such a line would be

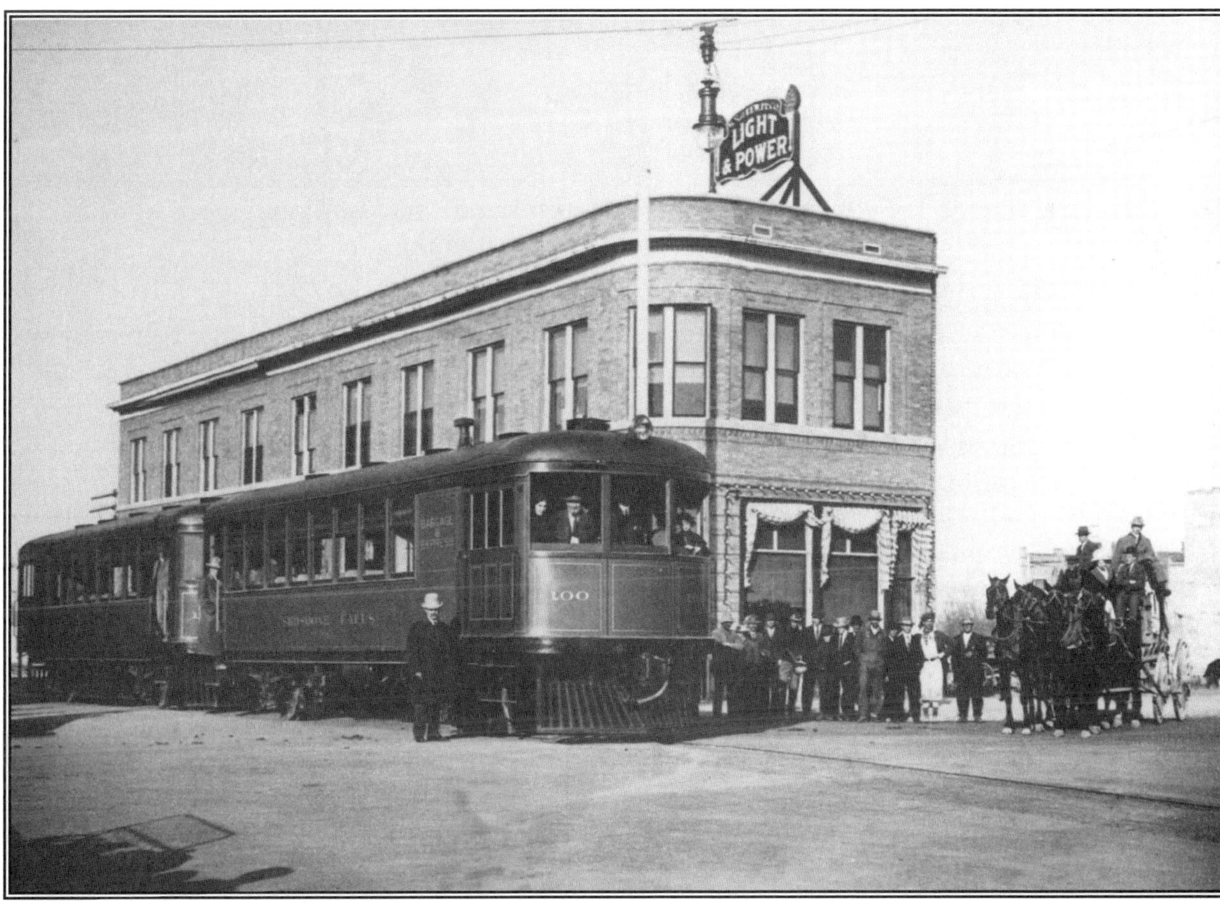

Battery-powered cars, at the power company building, 163 Main Ave. N., in May, 1913.

built to nearly all the towns and points of interest in Magic Valley, between Shoshone on the north, and Wells, Nevada on the south.

In 1909, Thomas A. Edison, the famous inventor, completed development of a battery that used an alkaline solution, rather than acid, for the electrolyte, and nickel and iron electrodes instead of the usual lead type. Subsequently, Edison built a light-weight streetcar, designed to use the battery, and he formed companies to manufacture both the batteries and cars.

Electric cars at rear of T. Falls High School, at the close of a school day in December 1913.
The county jail now occupies part of this site.

Perrine became taken with this idea of powering electric rail cars using storage batteries, rather than by conventional means. Sometime in early 1912, he obtained a franchise from the Twin Falls city fathers to build and run an electric line, and he, H. L. Hollister, and seven local men formed the Twin Falls Railway Company. Perrine became president, and his brother, Ledyard, was named chief engineer.

In May, Burt rode one of Edison's battery streetcars on a demonstration trip out of New York City, over Pennsylvania Railroad tracks. Later that month his company ordered two of Edison's cars and—from the Illinois Steel Company—a large number of rails.

Grading for a standard-gauge railway was begun within the city limits. Plans called for the electric line to originate at the Oregon Short Line (O. S. L.) depot, then cross Shoshone street, and run along Second (now Gooding) Streets West and North to the North Five Points. From there the tracks would make a ten mile loop—out Blue Lakes Boulevard to the canyon rim, thence east to Shoshone Falls; then south for several miles, and finally west to the city again. A battery-charging station would be placed at the corner of Second Street and Fifth Avenue West.

Perrine had hoped that the line would be in operation by December 1, 1912, but there were delays in building the cars and obtaining rails. By year's end, track had only been laid from the depot to North Five Points.

Twin Falls businessmen help finish the electric line roadway in August, 1914.

Perrine had given a promissary note to Illinois Steel for the price of the rails, and had gained possession of most of them. But, when the last 11 carloads arrived, in late January, 1913, he was unable to pay the freight charges, so the rails were impounded, stored, and later sold by the O.S. L. Failure to obtain the rails, and Perrine's growing financial troubles forced a decision to lay only the Addison Avenue leg of the loop. But little work was done.

When the storage battery cars finally arrived on April 29, people flocked to the rail yard to see them unloaded. The cars, numbered 100 and 101, were reportedly finished in yellow, with green trimmings. Each could seat 40 passengers, with space in the front end for baggage, a smoking compartment and the motorman's controls.

Perrine had a agreement with the O. S. L. for the interchange of traffic so, on Saturday evening, May 2, the cars made a trial run to Buhl, over the Short Line tracks, carrying 80 guests. The run was made in 35 minutes, each way, under the control of a representative from the factory. The cars were said to be commodious, well-ventilated and easy-riding.

With the arrival of the cars, the city council gave Perrine the first of several extensions of his contract time for completion of the line. Perrine promised that work would be resumed on the Addison leg, and he claimed that an amusement park owner planned to establish such a facility at some point on the electric line.

But, despite Perrine's assurances, no track was laid until October, 1913 when enough money was obtained to allow completion of the Addison Avenue leg to a point 3 miles east of town. About December 1 the cars began running there every weekday, morning and afternoon, carrying about 30 school children and an occasional adult passenger. Fares varied from five to twenty cents, depending upon the distance traveled.

Trolley cars at Shoshone Falls "station" in late 1914.

The Addison leg was finally completed to Shoshone Falls at the end of August,1914, but only after a committee of citizens gathered donations to help finish the track. With electrician-motorman C. M. David at the controls, a regular schedule was begun. Five round trips per day were made on weekends, and three each weekday.

And so, in this fashion, the trolley ran until July 1916, when operations ceased. In mid-September trips were made for a few days, carrying school children and workmen who were "fixing up the line." It was reported that Ned McCracken—head clerk at the Hotel Rogerson—had leased the system, and intended running a gasoline-powered truck on the rails.

But, apparently, the line was never used again. The note to Illinois Steel had never been paid; there were other liens against the railway property, all of which was eventually sold to satisfy creditors. A considerable portion of Perrine's personal assets were also sold to satisfy the company's debts.

The trolley project was one of the few times that Perrine's legendary vision proved faulty. He failed to recognize that neither the local population nor the tourist trade was large enough to support such a transportation system. He also under estimated the impact of the rapid growth in automobile usage.

And finally, Perrine didn't take into account the limitations of using battery power. Although claims were made that battery cars had also been sold to trolley lines in New York and Cuba, evidence suggests that Perrine's line was the first and only one to ever use them.

The Wizard of Menlo Park

It isn't known just how close Burt Perrine and Thomas Alva Edison became but, ideologically, they were perfect soul mates.

It was Edison who once declared "everything comes to him who hustles while he waits." The inventor often worked 18-hour days, catching occasional naps on cots that were strategically positioned in his large research laboratories at Menlo Park and West Orange, New Jersey.

In 1912, the year that Perrine ordered his battery-operated cars, Edison was hard at work developing a new phonograph, with a diamond needle, that recorded to a disc rather than a cylinder. That year, Edison would also produce the first talking motion pictures by synchronizing this phonograph with the projecting kinetoscope he had invented in 1888.

On one of his visits to the East, Perrine brought Edison several varieties of his Blue Lakes apples and, in return Edison sent Perrine one of the handsome new mahogany-encased Diamond Disk phonographs.

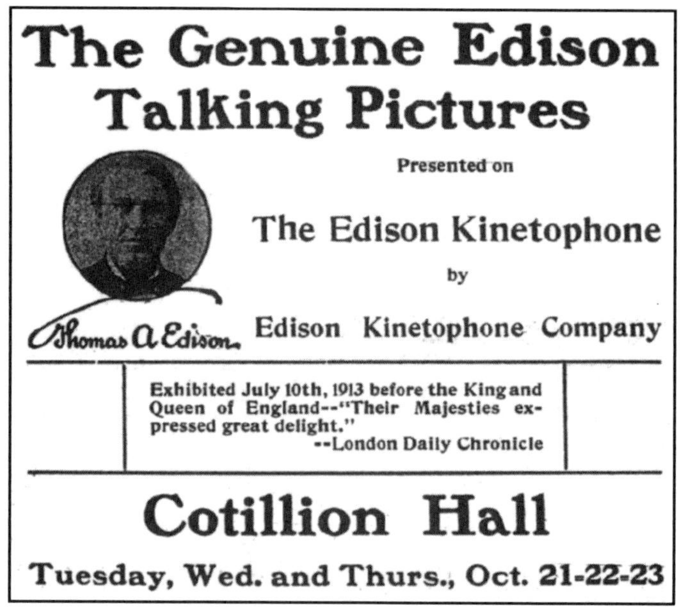

Ad from Twin Falls Times in October, 1914
(Author's collection)

Select Bibliography

Books, Monographs, Pamphlets and Articles

Bancroft, Hubert Howe. *History of Nevada, Colorado and Wyoming, 1540-1888*. San Francisco, 1890.

________. *History of Washington, Idaho and Montana, 1845-1889*. San Francisco, 1890.

________. *History of Oregon*, 2 vols. San Francisco, 1888.

Bolton, Chris. "Twin Falls and the Silver Screen," Typescript, Twin Falls Public Library, 2002.

Cherny, Robert W. *A Righteous Cause, The Life of William Jennings Bryan*. Boston: Little, Brown and Company, 1985.

"Clark's Ferry." *Idaho State Historical Society Reference Series.* No. 750, 1952.

Coleman, Kathleen. "The Cattle and Sheepmen Feud; the Story of Diamondfield Jack," *Albion State Normal Bulletin*, March 1941.

Cross, Major Osborne. A report, in the form of a journal, to the quartermaster general of the march of the regiment of mounted riflemen to Oregon, from May 10 to Oct 5, 1849, Washington, D. C., 1850.

Deed Book #1, Twin Falls County Recorders' Office.

Estes, Virginia, ed. *A Pause For Reflection*, Burley Idaho: Daughters of Utah Pioneers, 1977.

A Folk History of Twin Falls County. Twin Falls, Idaho: The History Publication Committee, comp., 1962

Foote, Mary Hallock, *A Victorian Gentlewoman in the Far West; the Reminiscences of Mary Hallock Foote*, ed. by Rodman W. Paul. San Marino, California: The Huntington Library, 1972.

Fuller, Dr. O. A. "Notes on the Twin Falls Municipal Band. Typescript. Twin Falls Public Library, undated.

Grover, David H. *Diamondfield Jack; A Study in Frontier Justice*. Reno, Nevada: University of Nevada Press, 1968.

Hailey, John. *The History of Idaho*. Boise, 1910.

Hayes, Anna H. " A Brief Summary Regarding the work and service of John E. Hayes in Twin Falls, Idaho." Typescript, Twin Falls Public Library, 1967.

________. "How a sage brush became a Christmas Tree." Typescript, Twin. Falls Public Library, undated.

________. "Town Sites on the Tract." Handwritten Ms., T Falls Public Library, undated.

________. "The Snake River Desert." Typescript, T. Falls Public library, 1967.

Hayes, John E. "Note by John E. Hayes." Photocopy, T. Falls Public Library. undated.

Holloway, James L. *Twin Falls County Pioneers*, Twin Falls Public Library, 2000.

Idaho Power Company. *Idaho Power Company; a History of the development of the Electric Industry in Southern Idaho, 1887-1943.* Boise, 1943.

Idaho State Historical Society Library, Mss 114, Box #9, Financial File & Box #14, Contracts and Deeds

Johnson, Claudius O. *Borah of Idaho.* Seattle: University of Washington Press, 1936

Kingsbury, H. J. *Bucking The Tide*. New York: Ganis & Harris, 1949.

Kowrach, Edward J. ed., *Journal of a Catholic Bishop on the Oregon Trail: The Overland Crossing of the Rt. Rev. A.M.A. Blanchet, Bishop of Walla Walla, from Montreal to Oregon Territory, March 23, 1847 to January 23, 1851.* Fairfield, Washington: Ye Galleon Press, 1978.

Laws of the Territory of Idaho, Third Session. Boise, Idaho, 1866.

Langford, Nathaniel Pitt. *Vigilante Days and Ways; The Pioneers of the Rockies."* 1890. Reprint. Freeport, New York: Books for Libraries Press, 1971.

Lovell, Edith H. *Benjamin Bonneville; Soldier of the American Frontier*. Bountiful Utah: Horizon Publishers, 1992.

Maddock, Robert K. Jr. "John Edward Hayes." Typescript. Author's collection, 2001.

________. "John Frederick Hansen." Typescript. Author's collection, 2001.

________. "Anna Hansen Hayes." Typescript. Author's collection, 2001.

Manning, Thomas G. *Government in Science: The U. S. Geological Survey, 1867-1894*. Univ. of Kentucky Press. 1967.

Moon, J. Howard. *A History of the Twin Fall Canal Company*. Typescript. Twin Falls Canal Company, 1985.

________, comp. *Twin Falls Canal Company Records.* Typescript, Twin Falls Public Library, undated.

Perjol, Tom. *The Frank H. Buhl Story*. Typescript. F. H. Buhl Trustees, 1999. Copy at T. Falls Canal Company.

Perkins, T. A. Asst. Engineer, Division No. 3, Twin Falls Land & Water Company. *Diary, 1903.* Photocopy, Twin Falls Public Library.

Preuss, Charles. *Exploring With Fremont; The Private Diaries of Charles Preuss, Cartographer for John C. Fremont*

on His First, Second, and Fourth Expeditions of the Far West. Transl. and Edited by Erwin G. and Elisabeth K. Gudde. Norman: Univ of Okla. Press, 1958.
Richardson, Albert D. *Beyond the Mississippi; From the Great River to the Great Basin. Life and Adventure on the Prairies, Mountains, and Pacific Coast.* Hartford, Connecticut, 1867.
Ricketts, Virginia. *Then and Now in Southern Idaho*, Jerome, Idaho: Falls City Publishing, 1998.
Roberts, Edwards. *Shoshone and Other Western Wonders*. New York, 1888.
Stacy, Susan M. *Legacy of Light; A History of Idaho Power Company*. Boise, Idaho: Idaho Power Company, 1991.
Strahorn, Carrie Adell. *Fifteen Thousand Miles by Stage.* (2 vols) 1911; Reprint. Lincoln: University of Nebraska Press, 1988.
The First Hundred Years, The History of Twin Falls, 1811-1911. Twin Falls High School Sophomore Class of 1944, Typescript.
Twin Falls City Directory, 1910. Twin Falls Public Library.
Twin Falls County Miscellaneous Record Book # 1, Twin Falls County Recorder's Office.
Twin Falls Canal Company Ditch Writer, Fall 2000, Twin Falls Canal Company.
"Twin Falls Fire Department Past and Present," Typescript. Twin Falls Public Library, Undated.
Twin Falls Village Council Minutes, 1907-1909. Typescript. Twin Falls Public Library.
Walgamott, Charles S. *Reminiscences of Early Days; A Series of Historical Sketches and Happenings in the Early Days of Snake River Valley*, Vol. 1, 1926; Vol. 2, 1927.
________. *Six Decades Back*, Caldwell, Idaho: The Caxton Printers, 1936.
Water Rights, Lincoln County, Book#1, Lincoln County Recorder's Office, Shoshone, Idaho.
Worster, Donald. *A River Running West; the Life of John Wesley Powell*. New York: Oxford Univ. Press, 2001.

Newspapers

Albion, Idaho *Albion Times*, 1903.
Boise *Idaho Daily Statesman*, 1867, 1877, 1885, 1890, 1897-1905, 1907, 1909, July 2, 1978.
Caldwell, Idaho *Tribune*, March 21, 1907.
Elko, Nevada *Independent*, 1870.
Gold Hill, Nevada *Daily Evening News*, 1864.
Hailey, Idaho *The Wood River News-Miner*, 1882-1884, 1886.
Idaho City, Idaho *World*, 1865.
Oakley, Idaho *The Weekly Times*, 1903.
Pocatello, Idaho *Tribune*, June 17, 1903
Portland, *Oregon Weekly Times*, June 19, 1858.
Portland, *Oregonian*, 1863, 1864.
Sacramento Union, 1864.
Salt Lake City *Herald* Feb 1, 1903; May 3, 1906.
Shoshone, Idaho *Shoshone Journal*, 1891–1892; 1896–1899; 1902-1904.
Silver City, Idaho *The Idaho/Owyhee Avalanche*, 1866, 1870, 1879, 1890, 1893.
Tuscarora, Nevada *Daily Times-Review*, 1881.
The Twin Falls Daily/Weekly News 1908-09, January 1918, October, 1927; February, 1928; August, 1929.
Twin Falls *Times*, November 11, 1904; April 1, 6, 1915, December 27, 1917.
Twin Falls *Times-News* January 25, 1954; October 3 & 5, 1943.
Unionville, Nevada *Humboldt Register*, 1868.

Index